W9-ANW-835

AMERICA'S MARINE SANCTUARIES

AMERICA'S MARINE SANCTUARIES

A PHOTOGRAPHIC EXPLORATION

NATIONAL MARINE SANCTUARY FOUNDATION

SMITHSONIAN BOOKS

WASHINGTON, DC

Published by Smithsonian Books
DIRECTOR: Carolyn Gleason
SENIOR EDITOR: Jaime Schwender
ASSISTANT EDITOR: Julie Huggins
EDITED BY Heather Dewar and Kristen Sarri
COPYEDITED BY Carla M. Borden
DESIGNED BY David Griffin

This book may be purchased for educational, business, or sales promotional use.
For information, please write: Special Markets Department, Smithsonian Books,
P.O. Box 37012, MRC 513, Washington, DC 20013

Library of Congress Cataloging-in-Publication Data

Names: National Marine Sanctuary Foundation (U.S.), author.
Title: America's marine sanctuaries : a photographic exploration / National Marine Sanctuary Foundation.
Description: Washington, D.C. : Smithsonian Books, 2020. | Includes bibliographical references and index.
Identifiers: LCCN 2019001825 | ISBN 9781588346667 (hardback)
Subjects: LCSH: National Marine Sanctuary System (U.S.) | Marine parks and reserves—United States. | Marine resources conservation—United States. | National Marine Sanctuary System (U.S.)—Pictorial works. | national marine sanctuaries, marine national monuments, and marine protected areas—United States—Pictorial works. | BISAC: SCIENCE / Earth Sciences / Oceanography. | NATURE / Ecosystems & Habitats / Oceans & Seas. | NATURE / Environmental Conservation & Protection.
Classification: LCC QH91.75.U6 N367 2020 | DDC 363.6/809162--dc23
LC record available at https://lccn.loc.gov/2019001825

Printed in China, not at government expense
24 23 22 21 20 1 2 3 4 5

PREVIOUS PAGES

A spectacular sunset glows behind surfers walking home along the wide beach at Neah Bay. The bone-chilling jade-green waters of Olympic Coast National Marine Sanctuary are home to incredible marine life and offer some great surfing locations.

FOLLOWING PAGES

A mother whale and her calf find refuge in the deep blue waters of Hawaiian Islands Humpback Whale National Marine Sanctuary. The sanctuary is one of the world's most important habitats for North Pacific humpback whales.

The sea is as near as we come to another world.

ANNE STEVENSON

The National Marine Sanctuary Foundation dedicates this book to all the people who helped establish our national marine sanctuaries and marine national monuments, the managers and volunteers who work tirelessly to protect them for future generations to enjoy, and the photographers who show us the wonders that lie beneath the waves and soar above the surface.

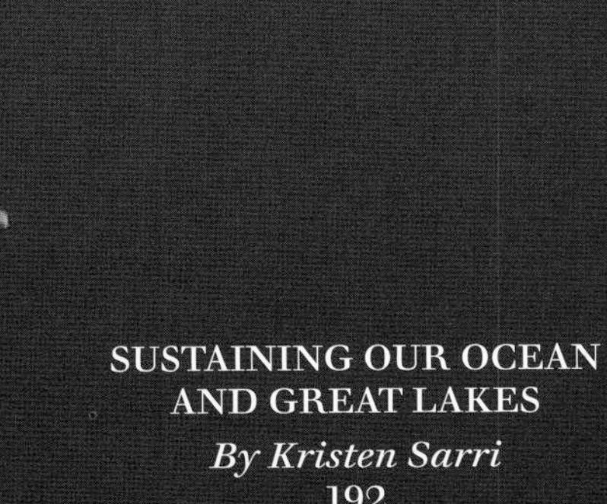

FOREWORD

BY SIGOURNEY WEAVER

MY LOVE OF THE OCEAN was a legacy my father passed on to me. Pat Weaver was a Navy man who had one requirement about where we lived: our home had to be in sight of saltwater at all times. Growing up, I listened to foghorns at night and was chased by crabs by day. I loved the mysteries of marine life. The ocean contains such diversity of life, and most of it is hidden from our sight. A lot of it is otherworldly to us, which makes the process of learning about the ocean and what lives below an unending series of surprises, a constant discovery of treasures. The ocean is full of organisms that are so unlike anything we know on land that their very existence seems impossible.

Our national marine sanctuaries and monuments are home to this incredible diversity of marine life. Papahānaumokuākea Marine National Monument, a sacred place to Native Hawaiians, protects one of the most remote archipelagos on our planet, and many of the species found there exist nowhere else on Earth. Adorable sea otters dive into the chilly kelp forests of Monterey Bay National Marine Sanctuary and emerge victorious, chomping on abalone clutched in their paws. Stellwagen Bank National Marine Sanctuary off the coast of Massachusetts is home to sixty-ton humpback whales that propel themselves out of the water to the enjoyment of people watching from boats. And the turquoise blue waters of the Florida Keys National Marine Sanctuary are the refuge of endangered and threatened corals that were once plentiful along the coast.

Watching waves crash over rocky shores and hearing the pounding of the surf, we are awed by the ocean's power. Looking out over a calm sea as the sun drops below the horizon, we find peace of mind in its tranquility. These features, which magnify the ocean's mystery and otherworldliness, can work to its disadvantage, because for many of us, the ocean's problems are out-of-sight and out-of-mind. Its vastness and power make it seem indestructible, with endlessly renewing resources. We forget that the ocean is both finite and vulnerable, and that we all depend on it for our survival, regardless of where we live or what we eat. Organisms in the ocean generate most of our oxygen, the ocean regulates our climate, and it provides a large portion

A Hawaiian monk seal rests with a green sea turtle on a beautiful beach in Papahānaumokuākea Marine National Monument. The marine monument offers sanctuary to both species, which are protected by the US Endangered Species Act.

of the world's population with sustenance. We cannot prosper unless the ocean prospers too. And the ocean is not prospering. Marine debris litters coastlines, entangles and kills wildlife, and damages underwater environments. Overfishing and illegal and unregulated fishing deplete species across the globe. Climate change acidifies and warms waters, killing coral reefs and making places uninhabitable for some wildlife.

We know that the ocean is not too big to fail. I hope we can see that it is too important not to defend. We know that when we protect nature, wild places and wildlife rebound. Scientists, conservationists, and world leaders are challenging us to protect at least 30 percent of the ocean by 2030 for the good of the planet and for the health and well-being of people across the world. Today, only 7 percent of our ocean is protected. There is more ahead for us to do as a nation and a world that relies on one global ocean.

Through the pages in this book, I invite you to discover the wonders of our National Marine Sanctuary System. To explore these protected waters through the eyes of those who care for them, and to learn how we can be stewards of these unique places. We rely on the ocean for life. We are its guardians. Join me. ○

1

THE OCEAN PLANET

The world below the brine,

Forests at the bottom of the sea, the branches and leaves,

Sea-lettuce, vast lichens, strange flowers and seeds,
the thick tangle, openings, and pink turf,

Different colors, pale gray and green, purple, white,
and gold, the play of light through the water . . .

The sperm-whale at the surface blowing air and spray,
or disporting with his flukes,

The leaden-eyed shark, the walrus, the turtle, the hairy
sea-leopard, and the sting-ray.

WALT WHITMAN

THE OCEAN IS ASWIRL WITH MOVEMENT. The seas that cover more than 70 percent of our blue planet are never still—they spiral around the world's ocean basins in five great gyres created by Earth's rotation. Every day, the gravitational force of the moon pulls the ocean back and forth across the globe. On a winter's morning, a storm off the coast of Indonesia can churn up giant swells that roll east, traveling halfway around the world to curl into perfect tubes and break on a California beach. On calm summer afternoons, saltwater flows through the spurs and grooves of a Florida coral reef, marking the reef crest with the faintest smear of foam.

The ocean—where 99 percent of all livable space on Earth exists—is a wondrous world full of activity. The reef hums with sounds: the clackety-clack of snapping shrimp, soldierfishes' drumlike thumps, the grinding of parrotfishes' powerful jaws as they feed on old, stony corals. On the seabed, spiny lobsters shimmy back and forth. To the surface, microscopic marine plankton whirl. Most forms of ocean life are in motion, at least for part of their lives, looking for food, mates, or safe havens. Sea urchins move in herds through the kelp forests of the Pacific Coast, grazing on dense stands of macro-algae. Sea otters forage in these kelp beds, helping to maintain the balance of nature in these complex underwater ecosystems. Great white sharks glide along the Atlantic seaboard,

An opalescent nudibranch inches across the brightly colored shell of a jeweled-top snail. These are two of the many incredible species to see in the tide pools of Monterey Bay National Marine Sanctuary.

PREVIOUS PAGES

Papahānaumokuākea Marine National Monument is one of the largest marine conservation areas in the world. Its name, Papahānaumokuākea, commemorates the union of the two Hawaiian ancestors—Papahānaumoku and Wākea—who gave rise to the Hawaiian Archipelago, the taro plant, and the Hawaiian people.

NATIONAL MARINE SANCTUARY SYSTEM

The National Marine Sanctuary System protects more than 600,000 square miles of our ocean and Great Lakes, from the South Pacific to the North Atlantic.

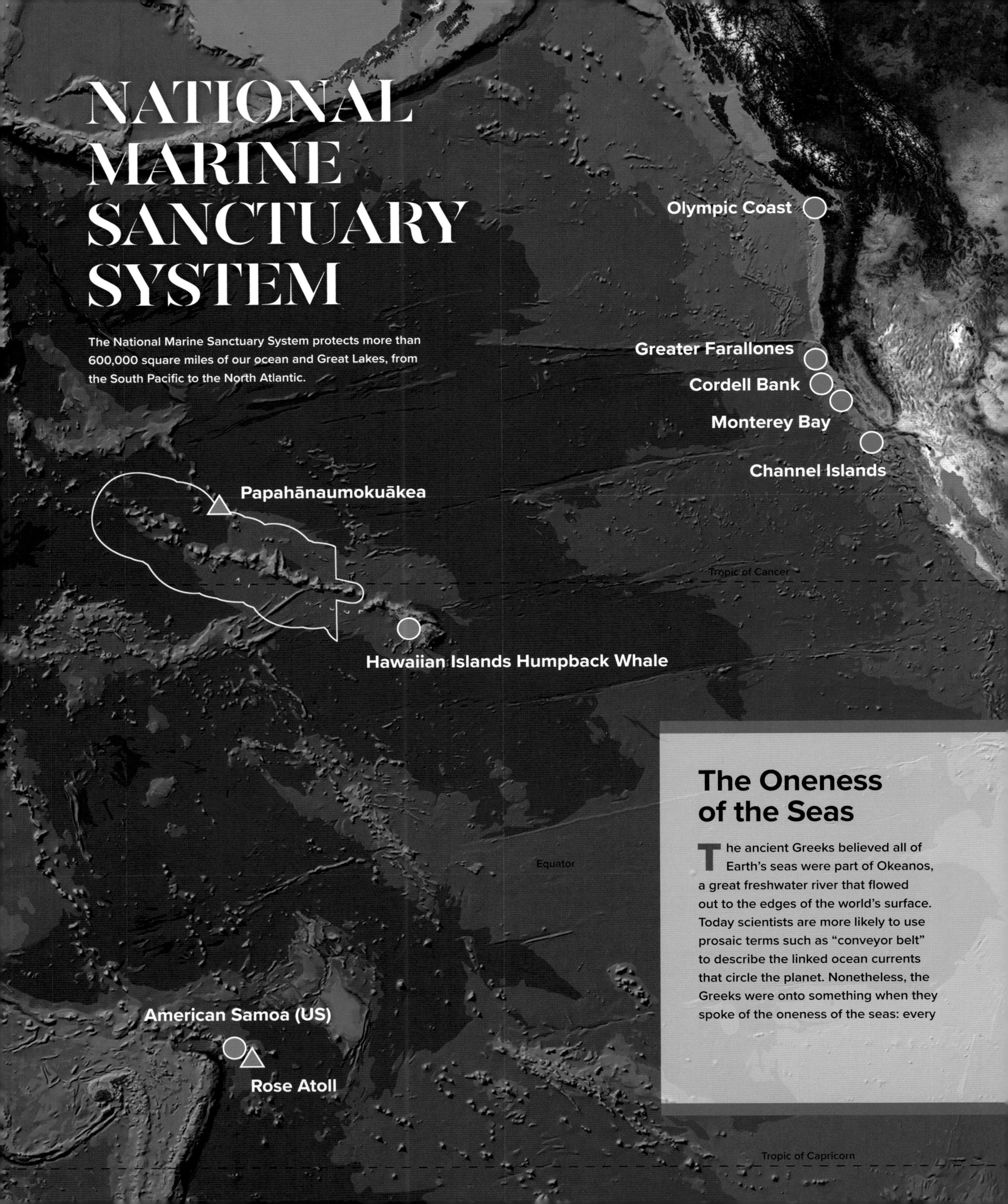

The Oneness of the Seas

The ancient Greeks believed all of Earth's seas were part of Okeanos, a great freshwater river that flowed out to the edges of the world's surface. Today scientists are more likely to use prosaic terms such as "conveyor belt" to describe the linked ocean currents that circle the planet. Nonetheless, the Greeks were onto something when they spoke of the oneness of the seas: every

part of the ocean is connected to the same whole. Human-drawn boundaries give rise to five ocean basins: the Arctic, Atlantic, Indian, Pacific, and Southern Oceans. They are delineated on maps, but nothing physically separates these waters. The Sea of Cortez is part of the Pacific Ocean, the Sargasso Sea is within the Atlantic, and each ocean basin flows into the other.

wintering off Florida and summering between Massachusetts and Nova Scotia. Young bluefin tuna born in the Sea of Japan travel more than 5,000 miles to California's coast to feed and grow to adulthood among abundant natural fish nurseries before traveling back to their birthplace to spawn.

Scattered across US waters, from the South Pacific to the North Atlantic, are more than a dozen special places known as national marine sanctuaries. Sanctuaries, like our national parks, are refuge and home to an immense array of wildlife and habitats. They preserve maritime and cultural resources that tell the history of our nation's past. In sanctuaries, natural forces, geographic features, seasonal patterns, plants and animals, historical artifacts, cultural practices, and human uses come together to create an ecosystem of extraordinary complexity.

Sea turtles have roamed Earth's oceans and seas for millions of years. There are seven species of marine turtles, five of which, including this hawksbill turtle, are found in the waters off the Florida Keys. All seven species of marine turtles are listed as either threatened or endangered under the Endangered Species Act.

Since the creation of the first national marine sanctuary in 1975, the United States has protected these special places for their natural beauty, ecological importance, and cultural significance, while still allowing people to use and enjoy them. The waters of the National Marine Sanctuary System encompass living organisms, the ocean and its currents, and the seafloor itself—which may be as flat as a prairie, carved into ravines deeper than the Grand Canyon, or pierced by mountain peaks taller than Mount Everest. Sanctuary waters provide protected habitat for marine species at risk of extinction and preserve historic shipwrecks and sacred cultural sites. They serve as outdoor classrooms for school and university students and living laboratories for marine scientists. They support important industries, including recreation, fishing and shellfishing, and tourism. They are places where people come to spend time with

The islands scattered across Papahānaumokuākea Marine National Monument are nesting sites for more than 40 different species of seabirds, including about 95 percent of the world's remaining population of the black-footed albatross, a relatively small and rare species.

At Journey's End, a Sanctuary

Some of the world's most intrepid long-distance travelers are ocean dwellers. Black-footed albatross roam the Pacific Ocean as solitary flyers for more than half the year, using a technique called dynamic soaring to ride the ocean winds on their six-foot wingspans. They almost never touch land: they feed mostly on squid and flying fish eggs, and have no need for freshwater because of a built-in seawater desalination system in their heads and beaks. But in winter, these birds reunite with their lifelong mates on remote patches of land in Papahānaumokuākea Marine National Monument, such as Midway Atoll National Wildlife Refuge, Kure Atoll, and Laysan Island, where they raise their young—a single chick per pair.

Albatross pairs rely on each other to ensure their chicks' survival and their own. One parent goes without food to stay on the nest with the egg or hatchling, while the other may fly thousands of miles to fill its gullet with food before returning to feed the chick and take over the nest watch. A favorite feeding ground is Cordell Bank National Marine Sanctuary, off the Northern California coast, where the dusky-feathered seabirds are abundant in springtime. The round trip from the monument to Cordell Bank is about 5,200 miles, a little bit farther than a round-trip flight from Portland, Maine, to Los Angeles, California.

The black-footed albatross had little sanctuary in the 19th century. Before the invention of chemical fertilizers, seabird droppings, which accumulated in vast quantities at nesting sites, were a valuable source of nitrogen and phosphorus for crops. European and American guano hunters raked seabird rookeries bare, carrying off the droppings for fertilizer and the birds' eggs for food, often shooting adult birds or destroying their nests. Populations were decimated. Recognizing the harm humans were causing to migratory birds, in 1918 the United States enacted the Migratory Bird Treaty Act to protect birds from these and other harmful actions. The federal law saved numerous species from extinction.

Kelp, a type of seaweed or marine algae, can form dense patches resembling an underwater forest. Channel Islands National Marine Sanctuary's kelp forests harbor a diversity of life with over 1,000 species of marine plants and animals.

families and friends and to fish, swim, dive, photograph, wade, explore, and enjoy the wonder of the natural world.

The United States is a maritime nation. It has one of the longest coastlines on the planet, at more than 95,000 miles. Its exclusive economic zone—the 200-mile-wide swath of ocean waters where each nation has special rights guaranteed by international law—is the second largest in the world. Throughout our nation's existence, the ocean has been essential for subsistence, commerce, and security.

The first legally reserved aquatic preserves were created by North American settlers in the coastal waters of the Massachusetts Bay Colony. The 1641 Body of Liberties, the first legal code established in New England, affirmed that every white male householder had the right to "free fishing and fowling in any great ponds and Bayes, Coves and Rivers, so farre as the sea ebbes and flowes within the presincts of the towne where they dwell." While early laws established rights to use and exploit the ocean, protecting marine areas in the United States for wildlife followed much later and often in response to overexploitation of a resource. One of the earliest marine conservation preserves created was in Alaska in 1869. Because the region's booming fur trade decimated seal populations, the federal government placed two of Alaska's Pribilof Islands and their surrounding waters off limits to all but Unangan subsistence hunters.

Until the early 20th century, underwater parks were essentially seaward extensions of protected areas on land, including state parks, national wildlife refuges, and national parks and monuments. Then in 1929, California established the San Diego Marine Life Refuge, offshore from the Scripps Institution of Oceanography. Two years later, early conservationist Julia B. Platt helped establish Hopkins Marine Life Refuge in Pacific Grove, California, seaward of Stanford University's marine research station on Monterey Bay. Platt was a neurobiologist who turned to civic activism after she could not find a suitable job in academia. She became the first mayor of Pacific Grove, and a promontory in the refuge is named in her honor. These two California marine refuges may be the earliest undersea reserves in the United States not created as afterthoughts to an adjacent park on land. Decades later, in 1959, Florida's John Pennekamp Coral Reef State Park was created specifically to protect the fragile coral reefs off Key Largo, Florida.

In 1968, a single photograph gave humankind an entirely new perspective on the delicate, ocean-ringed orb that is our shared home. On December 22 of that year, Apollo 8 circumnavigated the moon. For the first time, the astronauts captured a view of brightly colored Earth rising from the blackness of space at the dawn of a new day. The perspective was startling. From space, the world appears not mainly green but a brilliant, living blue. This is our ocean planet. The picture, considered one of the greatest environmental photographs ever taken, showed how fragile our planet is, and spurred people across the

William Anders, an astronaut on Apollo 8, captured this view of a brightly colored Earth rising from the blackness of space. This is our blue planet, our ocean planet.

globe into action to protect it. "Standing on the moon looking back at Earth—this lovely place you just came from—you see all the colors, and you know what they represent," astronaut Buzz Aldrin said of a later Apollo mission. "Having left the water planet, with all that water brings to the Earth in terms of color and abundant life, the absence of water and atmosphere on the desolate surface of the moon gives rise to a stark contrast."

The 1960s brought a wave of support for executive and legislative action to protect the ocean. In 1966, a science advisory committee to President Lyndon B. Johnson recommended the creation of a national marine wilderness preserve system. That same year, Congress enacted the Marine Resources and Engineering Development Act. The act focused attention on the nation's coasts and ocean and created the Federal Commission on Marine Science, Engineering, and Resources. The commission released its 126 recommendations on January 9, 1969;

the report highlighted the importance of the ocean as a new American frontier for exploration, and the need to protect the nation's ocean and coasts from pollution and overexploitation. It called for an overhaul of federal ocean and coastal programs, including the creation of an overarching National Oceanic and Atmospheric Administration (NOAA) with responsibilities similar to NASA's role in space, and the enactment of laws to protect the ocean and its resources.

Just days later, an environmental catastrophe revealed the vulnerability of the nation's ocean and coasts. Beginning on January 28, 1969, and continuing for ten days, a blowout on an oil drilling platform six miles off the coast of Santa Barbara, California, spilled an estimated 3 million gallons of crude oil into the ocean. The spill shut down commercial fishing, fouled beaches, and killed countless dolphins, sea lions, seals, and more than 3,600 seabirds. At the time, the disaster was the largest oil spill in US history. Americans witnessed the oil-coated animals' suffering far too close for comfort, through live television coverage. The public reacted with a new intensity of environmental concern and activism. The event marked a turning point in the nation's conservation history. The activism ignited by the spill helped spur the first Earth Day, celebrated by 20 million Americans across the country, and built bipartisan support for legislation to protect the ocean.

In 1972 and 1973, Congress passed a series of landmark conservation laws and created the framework that has guided US ocean science and policy for the past 50 years. Among these bedrock environmental laws were the Marine Mammal Protection Act, the Coastal Zone Management Act, the Federal Water Pollution Control Act, the Endangered Species Act, and the Marine Protection, Research, and Sanctuaries Act. One century after establishing Yellowstone as the first national park, and eight years after President Johnson's Science Advisory Committee recommendation, Congress acknowledged the need for marine protected areas to conserve our ocean, and established the National Marine Sanctuary Program.

Today the United States has more than 1,000 marine protected areas of all sorts, from national marine sanctuaries to local areas set aside for conservation or research. Together they cover 1.2 million square miles, or 26 percent of the nation's waters. They range in size from South 239th Street Park Conservation Area—a sliver of tidelands and seawater off a street-end beach access park in Des Moines, Washington—to Papahānaumokuākea Marine National Monument, which is more than twice the size of Texas. Private entities and communities, states and territories, tribes, and the federal government, often working in partnership, manage these protected areas.

Marine protected areas provide long-term safeguarding for valued habitats, species, and features in our oceans, coasts, estuaries, and Great Lakes. Familiar examples in US waters include national marine sanctuaries, national parks, national wildlife refuges, and their state, local, and tribal equivalents. Some marine protected areas , called reserves, are "no-take" areas, where nothing can be

Visitors from around the world come to the Florida Keys to experience the islands and the turquoise waters that surround them. In Florida Keys National Marine Sanctuary, a network of buoys makes it possible for boaters to enjoy the waters without damaging the coral reef and other natural and cultural resources.

harvested, and all the marine life is protected. Marine reserves are rare in the United States, with just over 3 percent of US waters in these no-take areas.

The remaining 23 percent of protected marine area is managed for multiple uses and priorities, including commercial and recreational fishing, scientific research, and wildlife and habitat protection. Within these areas, managers seek to balance conservation of natural and cultural resources with sustainable use. By area, the vast majority of protected ocean in the United States is within the Pacific Islands. Over 1.1 million square miles are held in trust for current and future generations.

Although Congress established the National Marine Sanctuary Program in 1972, no obvious candidate for the first sanctuary emerged until the following year, when scientists from Duke University and the Massachusetts Institute of Technology discovered the wreckage of USS *Monitor*. The famous Civil War ironclad warship's remains lay on the seafloor 16 miles off the coast of Cape Hatteras, North Carolina. The new, untested marine sanctuary program seemed the best option for protecting the remote site, so the governor of North Carolina requested that NOAA designate it as a sanctuary. In January 1975, a circle of ocean one mile in diameter, with the wreck at its center, became Monitor National Marine Sanctuary.

On January 30, 1975, the wreck of the Civil War ironclad USS *Monitor* received federal protection as Monitor National Marine Sanctuary. It became the very first site in the National Marine Sanctuary System, which now includes 14 national marine sanctuaries and two marine national monuments.

Through the 1980s, the nation established six more sanctuaries in locations from American Samoa to California to coastal Georgia. Then three disastrous ship groundings on the Florida Keys' coral reefs inspired Congress to put stronger protections in place there in 1990. Florida Keys National Marine Sanctuary, established in November 1990, incorporated two existing

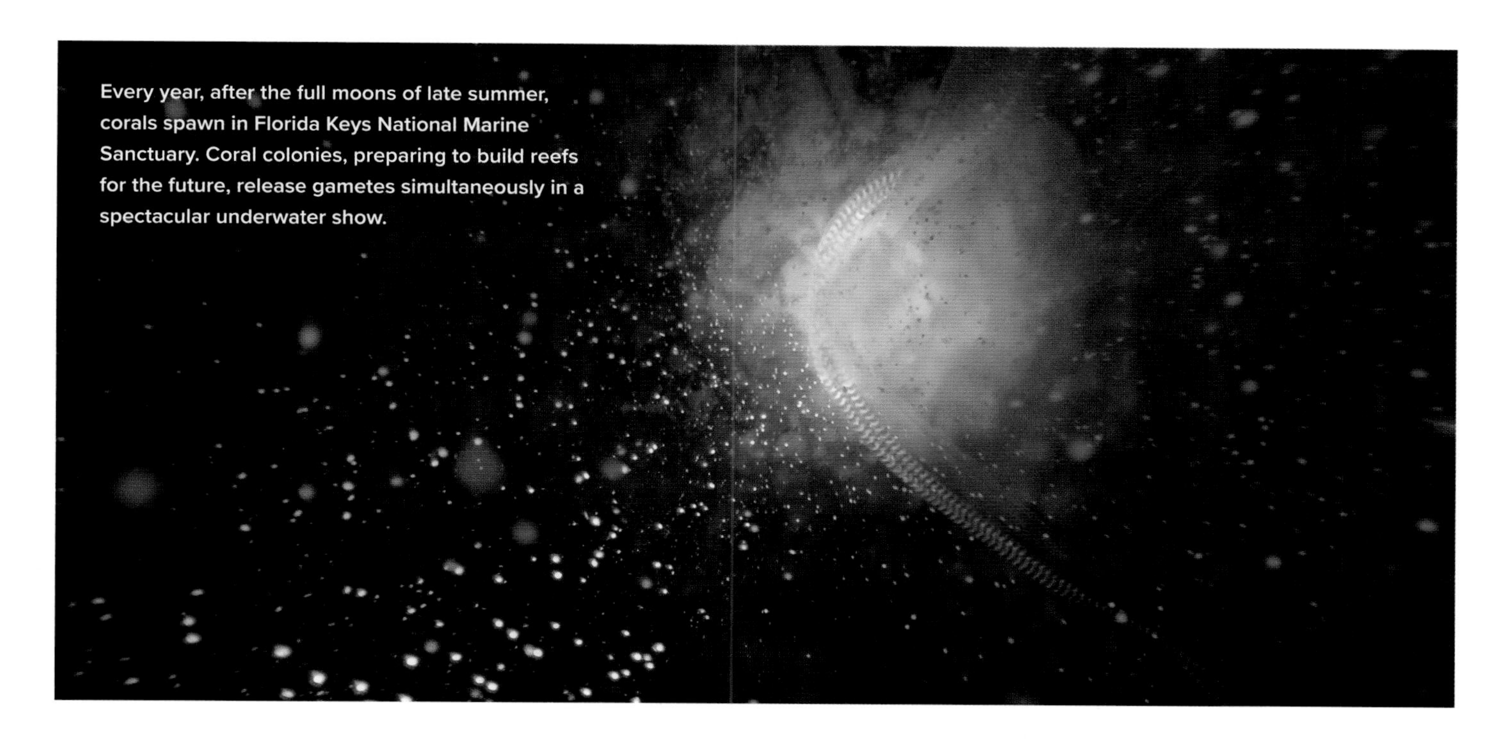

Every year, after the full moons of late summer, corals spawn in Florida Keys National Marine Sanctuary. Coral colonies, preparing to build reefs for the future, release gametes simultaneously in a spectacular underwater show.

Safe Havens for Seafarers

The ocean is open and unbound. Wayfaring sea creatures travel hundreds or even thousands of miles in search of safe places to feed and raise their young and may depend on habitats that are continents apart. Humpback whales spend their winters breeding in the Caribbean and their summers near Massachusetts's Stellwagen Bank. Sea turtles hatch on Atlantic beaches of the southern United States, such as Cape Hatteras and Cape Canaveral national seashores. They spend their early lives far out in the North Atlantic in the floating seagrass meadows of the internationally protected Sargasso Sea, before returning to lay their eggs on the beaches where their lives began. Whale sharks crisscross the Gulf of Mexico feeding on plankton and finding protected waters at both ends of their migration, in Flower Garden Banks National Marine Sanctuary off the Texas-Louisiana coast and in Mexico's Yum Balaam Nature Reserve.

Each summer, cold coastal currents carry swarms of small, nutritious sea creatures called krill down the Pacific Coast from Alaska through Canadian waters to central California. Blue whales, salmon, and other predators follow them. The three national marine sanctuaries on that part of the California coast protect all these creatures. The seafood buffet is especially abundant at Greater Farallones National Marine Sanctuary near San Francisco. Here, the cold coastal current meets warmer water rising from the seafloor, carrying nutrients that feed even more sea life. These waters support large populations of top predators, including powerful great white sharks.

Great white sharks make impressive, long-distance seasonal migrations each year. On the Atlantic Coast, some satellite-tagged great whites move from the waters off Massachusetts in summer to Florida Keys National Marine Sanctuary in winter and sometimes pass through sanctuaries as they go. In the Northeastern Pacific Ocean, great whites migrate between California, Mexico's Baja Peninsula, and the Hawaiian Islands, often stopping off at an area between Mexico and Hawai'i known as the "White Shark Café." This is the only known place where white sharks from Mexico and California regularly mingle. Its remoteness from land may make it an important haven for them as they travel between protected Mexican waters and the California and Hawaiian national marine sanctuaries.

Even tiny marine organisms may cross international borders. The Florida Keys lie on the downstream side of strong ocean currents connecting the western Caribbean Sea and the Gulf of Mexico with southern Florida. The Florida Current originates in the South Atlantic and the Caribbean Sea. It forms where the Gulf of Mexico Loop Current and the Yucatan Current come together and flow through the Straits of Florida and around the Florida Peninsula before joining the Gulf Stream. Scientists think this strong surface current carries warm water and larvae from the Caribbean to replenish the coral reefs and marine life of the Florida Keys.

Rose Atoll Marine National Monument, which is part of both the National Marine Sanctuary System and the National Wildlife Refuge System, is within the United States' only sanctuary south of the Equator, the National Marine Sanctuary of American Samoa.

The reef in Gray's Reef National Marine Sanctuary is not a typical one. Its foundation was formed from the consolidation of sediment, like shell fragments and sand, rather than the hard coral characteristic of tropical reefs.

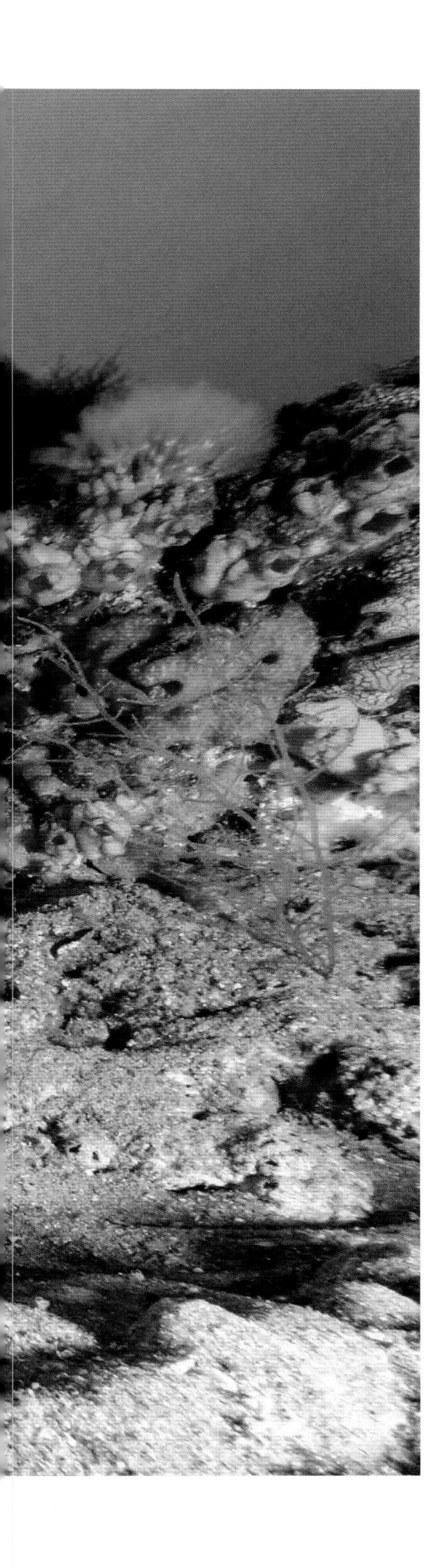

sanctuaries—the 134-square-mile sanctuary just off Key Largo with the smaller, exceptionally beautiful Looe Key sanctuary about five miles off Big Pine Key—and expanded the protected area. The new sanctuary encompassed 3,840 square miles of the waters offshore. The law immediately restricted ship traffic away from the reef and placed the sanctuary off limits to oil exploration, mining, and any other activity that might alter the seafloor.

By the end of the 1990s, the National Marine Sanctuary Program had grown to 12 sanctuaries, together embracing more than 9,500 square miles of ocean. In 2000 the nation designated the first Great Lakes sanctuary, Thunder Bay, in Lake Huron. The newest sanctuary is Mallows Bay–Potomac River National Marine Sanctuary, about 40 miles south of Washington, DC. Designated in 2019, it is the first sanctuary in the Chesapeake Bay watershed.

At this writing, there are 14 national marine sanctuaries and six marine national monuments, of which two—Papahānaumokuākea and Rose Atoll—are part of the National Marine Sanctuary System. Both monuments and sanctuaries protect America's underwater treasures, though they are created and administered differently. The National Marine Sanctuaries Act allows NOAA to identify, designate, and protect areas of the marine and Great Lakes environment with special national significance as national marine sanctuaries due to their conservation, recreational, ecological, historical, scientific, cultural, archaeological, educational, or aesthetic qualities. Congress may also create national marine sanctuaries in stand-alone legislation. The Antiquities Act of 1906 authorizes the US president to establish national monuments on federal lands, including in the ocean, that contain "historic landmarks, historic and prehistoric structures, and other objects of historic or scientific interest." The presidential proclamation that creates each national monument spells out how the federal government will manage and protect it. That is why some are part of the National Marine Sanctuary System and others are not. In 2006 President George W. Bush used the Antiquities Act to establish the nation's first marine national monument in the Northwestern Hawaiian Islands, creating the largest conservation area in the United States. President Barack Obama expanded the monument in 2016.

National marine sanctuaries and monuments protect marine wildlife and the habitats they call home. These treasured places are our essential network of protected waters. They connect us to our communities, our country, and our world. Each sanctuary protects habitats and maritime resources unlike any other on Earth. They are places where the public, communities, and businesses can engage in efforts to conserve our ocean and Great Lakes, for the good of the world and everything in it.

The following pages are an invitation to discover the wonders hidden under the waves of marine sanctuaries and monuments and to take part in protecting and conserving these extraordinary places, for the health of the ocean and of humankind. ○

2

WILDNESS AND WONDER

To swim in the ocean is to immerse myself in wildness,
to feel the way the water rises and falls like breath.

BONNIE TSUI

STRETCHING FROM THE CLEAR TROPICAL WATERS of American Samoa to the agate-green waves that break outside Boston Harbor, national marine sanctuaries reflect the infinite variety of the US ocean. Wind, light, currents, the shape of the seafloor, the mixture of warm and cold, of fresh and salt—these forces give each of the sanctuaries and monuments its unique character. They can seem unimaginably different from one another and from familiar earthly landscapes. Still, they are part of Earth's one ocean, connected by currents, wayfaring sea creatures, and the hands of humans, whose way of life shapes these watery places.

The Pacific

The United States' only sanctuary south of the Equator, National Marine Sanctuary of American Samoa conserves natural and cultural treasures. The strong, warm, counterclockwise currents of the South Pacific Ocean bathe the five volcanic islands and two coral atolls of American Samoa. The coral reefs here are healthy and varied, with more than 250 different coral species. The sanctuary is home to 1,400 types of marine invertebrate animals, including sea stars, sea urchins, and brightly colored giant clams, as well as sea turtles, humpback whales, dolphins, and more than 950 species of brilliantly hued fish. Scientists studying one 22-foot-tall mounding coral colony off the island of Ta'u estimate that it is about 500 years old. About 200 million individual coral animals make up the surface of this colony. This part of the sanctuary is home to several of these large Porites coral. It is rightfully called the Valley of the Giants. Within the sanctuary is Rose Atoll Marine National Monument. Shallow reefs are sprinkled with pink patches that look like rose-colored cement but are actually a type of seaweed called crustose coralline algae. These abundant pink crusts make the atoll's reefs distinct from those around other Samoan islands and give the monument-within-a-sanctuary its name.

This is Fale Bommie, or Big Momma, one of the largest corals in the world. Located in the Valley of the Giants in National Marine Sanctuary of American Samoa, this Porites coral has a circumference of 134 feet, stands 21 feet tall, and is more than 500 years old.

PREVIOUS PAGES

The kelp forest of Big Sur surrounds a gray whale and her calf. They are migrating through Monterey Bay National Marine Sanctuary from the warm-water lagoons of Baja, where the calf was born, to feeding grounds in Alaska. At least eight different species of whales and dolphins frequent the protected waters of the sanctuary.

Hawai'i is the only US state where humpback whales go to mate, calve, and nurse their young. Here a humpback whale and calf glide gracefully through the waters of Hawaiian Islands Humpback Whale National Marine Sanctuary.

The Way of the Whales

Humpback whales are famous for their acrobatic leaps and spins, their distinctive long flippers, and the eerie beauty of their songs. Long-lived, sociable, and once in danger of extinction, these great whales make a round-trip journey of thousands of miles from their Arctic Ocean summer feeding grounds to the warm tropical seas where they spend the winter. Most of the North Pacific population return each winter to Hawaiian waters.

Mothers nursing calves born in the previous year in Hawai'i are generally the first to arrive in November. Adult males follow and then adult females. Last to arrive are the pregnant females, who stayed longer in the Arctic to feed on krill, the shrimplike creatures that are one of the ocean's most abundant and important food sources. After a 12-month pregnancy, the females give birth in late winter. Whale watchers often see mothers and calves swimming together near shore.

The warm waters of Hawai'i are an ideal nursery for humpback whale calves, which are born without the thick blubber that insulates the adults from cold. Though full-grown humpbacks are about 40 to 45 feet long and nearly invulnerable to predators, a newborn is potential prey for orcas, which don't typically inhabit the tropical oceans. By the time mothers and calves depart for Alaska in late spring, the calves have grown to about 24 feet and learned to stay close to their mothers.

Human history in American Samoa dates back over 3,000 years. The sanctuary helps preserve historical artifacts, including pottery from the early Polynesian Lapita culture, whose people reached Samoa about 3,600 years ago. The sanctuary also preserves Polynesian plainware ceramics produced across the inhabited islands of the Samoas from 1,500 to 3,000 years ago.

When founded in 1986, National Marine Sanctuary of American Samoa was the smallest one in the network, a quarter-mile patch embracing a single bay. It expanded in 2012 to almost 13,600 square miles, nearly four times the size of Yellowstone National Park. It encompasses bays formed by collapsed volcanic craters, vibrant coral communities, open ocean, colossal seamounts, and unique fish and seabird populations. The sanctuary protects these places through regulations that ban sand mining, prevent the introduction of nonnative species, and prohibit boats from anchoring on the corals.

Polynesian voyagers from the South Pacific to Hawai'i followed the constellation known to Western astronomers as the Southern Cross and in the Hawaiian language as Hanaiakamalama, which translates to "cared for by the moon." Travelers crossed the Equator, then followed the North Star as they reached the lower edge of the North Pacific Gyre, which brushes the Hawaiian Archipelago. The islands and ring-shaped coral atolls of the archipelago are the exposed peaks of a 1,500-mile-long undersea mountain chain. The chain formed when a portion of the Pacific Plate, a large piece of Earth's crust, moved over a volcanic hot spot on the seafloor. The Big Island of Hawai'i, in the southeast, is the youngest island, at only about 400,000 years old; Kure Atoll, to the northwest, is an estimated 30 million years old.

The waters of Hawai'i are every shade of blue: turquoise, aquamarine, azure, peacock, cobalt. David Malo, a 19th-century historian and a Native Hawaiian, described the point where the sea meets the sky as *kūkulu-o-ka-lani*, the walls of heaven. In these tropical seas, flamboyant butterflyfish, angelfish, and triggerfish school over coral reefs, while brightly colored sponges, lobsters, crabs, shellfish, anemones, and sea stars live atop the reefs, in their crevices, and on the seabed. Green sea turtles graze on seagrass meadows and on sunny days sometimes rest alongside Hawaiian monk seals on sandy beaches. In winter and spring, albatross, boobies, shearwaters, and terns nest onshore. Galapagos sharks and tiger sharks patrol the open seas, making the ocean around the remote Northwestern Hawaiian Islands one of the last ecosystems on the planet dominated by top predators.

The main Hawaiian Islands lie at the heart of Hawaiian Islands Humpback Whale National Marine Sanctuary, the Pacific's most important nursery for humpback whales. Place names throughout the islands echo the animals' presence: both Kaua'i and the Big Island of Hawai'i have locations named Koholālele, or "leaping whale." Humpback whales were once plentiful in the ocean, but commercial whaling globally depleted their population.

Today a variety of laws protect humpback whales, including the Marine Mammal Protection Act, the Endangered Species Act, state wildlife laws, and the National Marine Sanctuaries Act. In 1992, to protect the animals and prevent habitat destruction, Congress passed a law to create the sanctuary, which stretches along the shorelines of Maui, Kaua'i, the North Shore of O'ahu, and the Kona and Kohala coast of Hawai'i island to an offshore contour line where the ocean depth reaches 600 feet.

Beyond the eight main islands of Hawai'i lies a seaway dotted with islets, atolls, and banks. The Kumulipo (the Hawaiian creation chant) describes the Hawaiian universe as comprised of two realms, Pō and Ao. Pō is a realm of deep, primordial darkness reserved for the gods and spirits, from which all life emerges and to which it returns after death. Ao is the realm of light and the living. Ke Ala Polohiwa a Kāne (the dark shining path of Kāne), also known as the Tropic of Cancer, is considered the border between Pō and Ao. The island of Mokumanamana is located on this boundary and is the center of convergence between the two realms; it sits near the entrance of Papahānaumokuākea Marine National Monument as only the second island in the northwestern part of the chain.

In homage to that tradition, the name Papahānaumokuākea Marine National Monument commemorates the union of Papahānaumoku and Wākea, the divine parents of the island chain, the taro plant, and the Hawaiian people. Some of the islands have several names: one or more Hawaiian names that highlight a natural feature such as an abundance of sharks, or a sacred quality ascribed to the place in traditional teachings, and an English name that often commemorates a historic shipwreck nearby. At more than 580,000 square miles, the monument is the largest conservation area in the country and one of the biggest in the world. It is also a United Nations Educational, Scientific and Cultural Organization (UNESCO) World Heritage Site, designated in 2010 for its outstanding value, both natural and cultural, to the heritage of humankind.

The Northwestern Hawaiian Islands are home to more than 7,000 marine species. These islands and atolls—Kure (Moku Pāpapa), Midway (Moku Pāpapa), Pearl and Hermes (Holoikauaua), Lisianski (Holoikauaua), Laysan (Kauō), Maro Reef (Ko'anako'a), Gardner Pinnacles ('Pūhāhonu), French Frigate Shoals (Kānemiloha'i), Necker Island (Mokumanamana), and Nihoa—provide breeding areas for Hawaiian monk seals and four species of sea turtles, nesting sites for more than 14 million seabirds, and more than 5,000 square miles of coral reefs.

Because this region is so remote—nearly 3,000 miles from the nearest continent—life forms evolved here that exist nowhere else on Earth. Researchers working in Papahānaumokuākea Marine National Monument continue to encounter new species: since 2000, scientists have discovered scores of new species of fish, coral, invertebrates, and even algae. Remarkably, on a 2015 expedition, scientists from NOAA and other institutions found that some deep

This day octopus (Hawaiian: *he'e*) in Papahānaumokuākea Marine National Monument is showing off its camouflage skills. Octopuses and other cephalopods have specialized cells called chromatophores that allow them to change their colors; they can also change the texture of their skin to mimic the environment around them. This camouflage helps them remain unseen by both predators and prey.

reefs in Papahānaumokuākea were inhabited only by endemic species, creatures found nowhere else in the world. This is the only known marine area in the world where all resident species are endemic.

In 2015, scientists, scholars, and Native Hawaiian cultural experts collaborated to develop scientific names and descriptions for newly discovered algal species, commonly called sea lettuce, that combine standard botanical terms based on the Latin language with Hawaiian words and concepts. The experts respected the Hawaiian tradition that naming something gives it life. Among the names chosen were two that honor important Hawaiian gods: Kū, the god of war, strength, and healing, and Kanaloa, the divine teacher of magic and an ocean deity. One species was named *Umbraulva kūaweuweukuaweuweu*, "grass of the god Kū." Another is *Umbraulva kāloakūlaukaloakulau*, meaning "the god Kanaloa alights upon the leaf."

At least 23 species protected under the US Endangered Species Act inhabit this marine national monument and the two national wildlife refuges and two state protected areas within its boundaries. For example, Papahānaumokuākea provides nearly the entire Hawaiian nesting habitat for the threatened green turtle. On the undisturbed beaches, the turtles come ashore to bask in daylight, a behavior no longer seen in most other parts of the world.

Not all the protected species are sea creatures. The Laysan duck was on the verge of extinction in the early 20th century, its population down to just 11 birds on a single island, when wildlife experts began working to save it. Today about 850 to 1,000 of these birds—still critically endangered—inhabit Laysan Island and Midway and Kure atoll. Midway, once a major US military base, is now a national wildlife refuge where millions of seabirds nest. Among them are the far-roving albatross; great frigate birds, known for the males' big throat pouches that inflate like balloons during courtship; and the rare, ethereal-looking white terns or fairy terns.

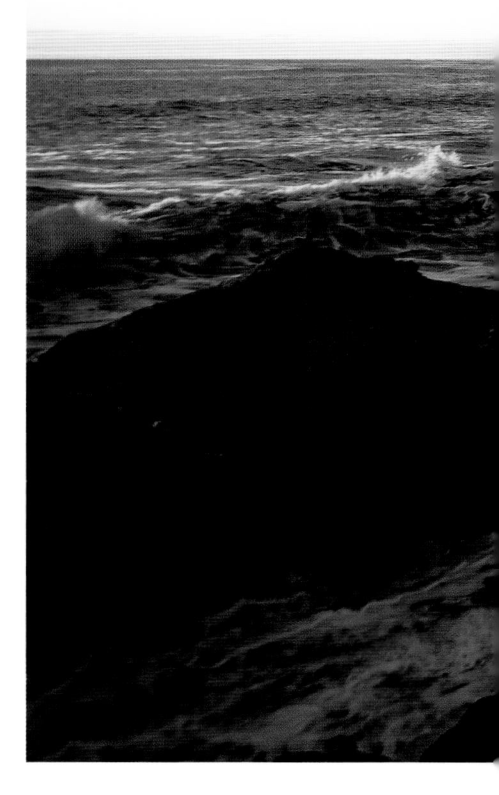

West Coast

The chilly California Current sweeps water southward from southern British Columbia in Canada to Baja California in Mexico, connecting five national marine sanctuaries that together protect some of the most diverse and productive cold-water areas in the United States. The current flows over lush cold-water corals growing on banks and seamounts, through drifting kelp forests and rocky intertidal zones, past underwater canyons and deep-sea vents.

North America's Pacific Coast is a geologically violent place. The planet's tectonic plates—the broken slabs of rock that make up Earth's crust—are in motion, and most of that grinding, slipping, fusing, and ripping takes place under the sea. Its continental shelf drops off suddenly, forming a ragged edge and submarine canyons, many originally formed on land when ancient rivers cut through rock and later drowned beneath rising seas. In the ocean's depths, the seafloor is constantly changing, pushed outward by molten rock rising in Earth's crust, or pulled down into the molten zone at deep-ocean trenches.

Near the zones where undersea tectonic plates meet, gaps in the seafloor give off methane and other gases. In these chemical stews, at depths where no light penetrates, creatures evolved to synthesize energy from gases instead of sunlight. There are at least 500 methane seeps along the Pacific Coast from Washington to California. Volcanoes also arise in these suture zones. Where they break the sea's surface, they form islands. As they go dormant and erode, coral atolls form on their sunken rims, like the Northwestern Hawaiian Islands, which lie atop drowned volcanoes.

Along the United States' Pacific Coast, cold ocean water rises up the slopes of the continental shelf and up submerged undersea mountains, or seamounts, lifting the detritus of long-dead sea life from the ocean floor. These organic sediments, rich in nutrients, feed a succession of sea creatures that begins with plankton and ends with tunas, sharks, whales, seabirds, and other top predators.

Schooner Gulch State Beach, located near Greater Farallones National Marine Sanctuary, contains over 50 acres of shores and cliffs to explore. The beach and headlands offer scenic spots along the coast for watching sunsets, and the beaches are popular for surfing, fishing, and picnicking.

Olympic Coast National Marine Sanctuary runs along 135 miles of Washington State's coast, much of it a wild, rugged shoreline that includes Olympic National Park and the coastal reservations of four sovereign governments: the Hoh, Makah, and Quileute tribes, and the Quinault Indian Nation. The sanctuary stretches up to 40 miles offshore, past sandy beaches interspersed with cliffs and pinnacles deeply eroded by the region's powerful winter storms. Gray whales swim offshore during their annual migration from the Arctic to warmer waters in Mexico, and harbor seals rest on the beaches. A total of 29 marine mammal species inhabit these waters for some part of the year. The cold waters teem with small fish that provide food for a variety of seabirds. The common murre can dive deeper than 240 feet; the rhinoceros auklet beats its wings underwater to "fly" in pursuit of a single fish; and the tufted puffin with its black and white feathers often uses its distinctive oversized orange beak to amass a catch of several small fishes. These birds and others cling to burrows, hollows, and crevices in the offshore sea stacks and islands, while the open waters and seafloor shelter salmon, rockfish, halibut, flounder, and cod.

Human presence on the Olympic Coast predates the historical record, and the artifacts that have endured to the present day attest to the long-standing relationship the tribes have with the marine environment on which they depend. In the mid-1800s, Isaac Stevens, governor and superintendent of Indian affairs of the Washington Territory, negotiated the Treaty of Olympia (1855) and the Treaty of Neah Bay (1855). Through these treaties, the Coastal Tribes ceded title to thousands of acres of land. In return, they were to receive reservation homelands for their exclusive use and were promised assistance from the United States. Importantly, the treaties reserved the rights of the Coastal Tribes to continue to hunt and gather resources at their usual and accustomed place to maintain their lifestyles and economies.

Olympic Coast National Marine Sanctuary's marine resources are managed through a unique, complex arrangement of overlapping federal, state, and tribal jurisdictions. The sanctuary lies within the Usual and Accustomed Treaty fishing areas recognized by treaty of the Quileute, Makah, and Hoh tribes and the Quinault Indian Nation. The Coastal Tribes are co-managers, with the state of Washington and the United States, of fisheries and related marine resources off the Olympic Coast. The sanctuary and the tribes act collaboratively to shape policy, research, and education programs, bringing different perspectives to the conservation of the priceless ecological and cultural legacy they share.

As the California Current moves south along the Pacific Coast, it carries seeds, spores, fish larvae, and adult animals toward the only place in the country where three sanctuaries connect to one another. This part of the ocean is famous for abundant sea life because of a phenomenon called seasonal upwelling: in summer, strong winds blow across the ocean's surface layers, pushing

Olympic Coast National Marine Sanctuary has an abundance of fish species. The red Irish lord is a type of fish known as a sculpin. Sculpins are benthic fish, living on the ocean floor.

waters offshore and causing colder waters to rise from below. These waters carry nutrients from the ocean depths which, when bathed with sunlight, spark a rich, productive food web. The nutrients are so plentiful and the ocean's productivity is so great that migratory populations of threatened and endangered sea turtles, fishes, seabirds, and whales travel thousands of miles to feed alongside creatures that live here year-round.

The most remote of these three special places is Cordell Bank National Marine Sanctuary, which sits 50 miles offshore, just northwest of San Francisco. Its focal point is a 40-square-mile granitic bank, once an island during the last Ice Age. The bank's shallowest peak is now 115 feet underwater, but it still acts as an underwater island of sorts—a garden spot above the seafloor, inhabited by cold-water coral and brilliant pink, orange, and yellow sea anemones, sea stars, and other solitary and mobile invertebrates.

The 1,286-square-mile sanctuary includes the bank, nearby Bodega Canyon, and a jumble of granite and rock reefs, boulders, cobbles, sand, and mud. Sponges, sea stars, sea cucumbers, and moon jellies filter their food from the nutrient-rich water. These creatures offer a buffet to a range of other animals, including tuna, humpback whales, Pacific white-sided dolphins, California sea

Cordell Bank sits at the edge of the continental shelf and rises sharply from its soft sediments to within 115 feet of the ocean surface. Here silvery fish swim over the strawberry-colored anemones, which help to give the bank its brilliant colors.

lions, and seabirds. More than 20 species of rockfish use the nooks and crannies of the bank as a nursery habitat during their first year of life. With the richness in the water, the waters throughout the Cordell Bank National Marine Sanctuary are a destination feeding area for pelagic seabirds. Up to five species of albatross have been seen there—the highest number of species of albatross seen north of the Equator. The steep walls of Bodega Canyon, the bottom of which lies over a mile deep, host abundant communities of deep-sea communities.

The bank was discovered by accident in 1853, when a crew of marine surveyors working off the California coast got lost in a fog. George Davidson of the US Coast Survey dropped a sounding line, expecting to find he was in about 400 feet of water. Instead, it read only about 180 feet. The survey vessel was over a bank—land rising from the ocean floor somewhere off Marin County—but exactly where, Davidson could not say. Sixteen years later, Captain Edward Cordell of the US Coast Survey was dispatched to find and document the bank. Cordell did what fishers the world over have done: knowing that seamounts are rich feeding grounds for sea creatures, he looked for a congregation of birds and whales. He rediscovered the bank in 1869, sounded it, and mapped it.

Edward Cordell also mapped Stellwagen Bank off Massachusetts, which later became a national marine sanctuary.

Then in 1978, physicist and scuba diver Robert Schmieder mounted a scientific expedition to dive on Cordell Bank, document its seascape and life forms, and study them. Schmieder and his nonprofit group, Cordell Expeditions, became the first group of divers to reach Cordell Bank. They found enormous schools of rockfish and lush undersea gardens of colorful sponges, sea anemones, sea stars, sea cucumbers, snails, and crabs. The group's work eventually helped lead to the bank's protection as a national marine sanctuary in 1989.

In 1978 an interviewer asked Schmieder to describe his first glimpse of the undersea plateau. "Surprisingly soon, I started seeing what I assumed was the bottom," Schmieder said. "It was sort of this greenish gray opaque cover below me. As I went down, this gray-green blanket started getting texture and a mottled appearance, and then I realized it was shimmering. And then, only then, did I realize I was looking at fish. I had not actually seen in my previous diving a solid opaque blanket of rockfish. And so I was unprepared to even recognize it. . . .They started slowly parting, as fish do. They moved very gently and slowly. . . . And if you can imagine having an opaque curtain in front of you and then a small hole opens and it widens like an iris. And as that iris opened I saw below me this extraordinarily colorful, exquisitely beautiful, astonishingly bright landscape. . . .[It] was an astonishing, overwhelming visual experience."

The exploration of Cordell Bank reaches depths and locations that remain mostly inaccessible to technical divers. Today a remotely operated vehicle, or ROV, guided by scientists on a research vessel at the surface, explores the depths. These scientists study the diverse communities on the bank, monitor them for changes, and explore new areas in the deepest parts of the sanctuary.

The waters of Cordell Bank merge into Greater Farallones National Marine Sanctuary, which stretches south and west from the rugged redwood coast of Mendocino to the Farallon Islands, a wilderness of rocky islands and sea stacks, and to the coastal lagoons and headlands near San Francisco's famed Golden Gate. About 400 fish species, from the flat-bodied halibut to the swift coho salmon, and hundreds of invertebrates, including urchins, sea anemones, red abalone, crabs, snails, squid, and sea stars, inhabit the sanctuary. Elegant snowy egrets wade in the coastal wetlands, brassy Caspian terns scoop fish from near-shore waters, and tens of thousands of sooty shearwaters fill the skies above the ocean.

Thirty-six marine mammal species, including one of the southernmost US populations of threatened Steller sea lions, find sanctuary in Greater Farallones. In fall, elephant seals and sea lions come to the Farallon Islands to breed,

Playful and loud, California sea lions are agile swimmers: they can "porpoise," or leap high out of water. They hunt offshore for fishes and squid. The population of sea lions has increased greatly since 1972, when hunting of marine mammals was banned in the United States.

and one of the planet's largest congregations of white sharks gathers to feed on them. Gray whales migrate along the shore, while blue and humpback whales throng from late spring through fall, when leatherback sea turtles may also visit. Harbor seals, California sea lions, and harbor porpoises feed, mate, raise their young, rest, and play in the sanctuary. Northern fur seals were hunted almost to extinction in the 19th century, but after an absence of 170 years, they have returned to breed in the Farallon Islands.

Seals gliding southward on the California Current slip imperceptibly into the waters of Monterey Bay National Marine Sanctuary, nicknamed the "Serengeti of the Sea" for its extraordinary abundance and variety of life. "Along the coast are great numbers of gulls, cormorants, crows, and other sea-fowl," wrote 18th-century Mexican historian Father Miguel Venegas of Monterey in *A Natural and Civil History of California*. "The sea abounds with oysters, lobsters, and crabs. Also huge sea wolves and whales." The Spanish explorers' sea wolves are known today as sea lions. Once in decline, sea lions and whales are now rebounding off modern Monterey, where oysters, lobsters, and crabs are still plentiful. Scientists report that more than 500 kinds of fish, at least 180 species of seabirds and shorebirds, and 36 marine mammal species use the sanctuary waters.

It is a place of rugged contrasts. Monterey Canyon lies more than two miles deep; it is more than a mile from the rim to the floor, the same size as the Grand Canyon. Underwater cameras reveal that it too has spectacular cliffs, spires, and valleys. Davidson Seamount, in the southern part of the sanctuary, is one of the largest known seamounts, rising 7,480 feet from its base to summit, and still it is more than 4,000 feet below the ocean's surface. Centuries-old sea fans, sponges, and deep-sea corals top the seamount, an extinct underwater volcano. In October 2018, a research expedition to Davidson Seamount discovered an "octopus garden," only the second such discovered in the world. A conservative estimate puts the population at 1,500 or more, possibly making this site an important nursery for these creatures. From its dense kelp forests to its boisterous elephant seal colonies, and from its soaring black-footed albatross to its enormous blue whales, Monterey Bay National Marine Sanctuary offers some of the most spectacular wildlife viewing in the world.

Near the base of California's rocky spine, off Santa Barbara, the cold California Current meets the Southern California Countercurrent, introducing warm water northward from Baja California. The warm and cold currents converge in Channel Islands National Marine Sanctuary, creating a unique blend of species and habitats around San Miguel, Santa Rosa, Santa Cruz, Anacapa, and Santa Barbara islands. Massive forests of cold-loving giant kelp shelter animals that usually prefer warmer Mexican waters, like spiny lobsters and moray eels. The sanctuary marks the northernmost range of the highly endangered white abalone. Colonies of pelicans, murrelets, cormorants, and guillemots breed on the islands and feed over the open water. Island beaches also host elephant

Channel Islands National Marine Sanctuary and National Park provide habitat for breeding populations for four species of pinnipeds, including the harbor seal. Harbor seals can dive to depths of 1,400 feet and remain underwater for nearly 30 minutes without resurfacing.

seals, fur seals, and sea lions. Humpback, blue, and sei whales swim offshore. A highly diverse array of habitats ranging from intertidal rocky habitat and sandy beaches to inshore kelp forests, soft bottom habitats, rocky reefs, and deep-sea coral gardens are found within the sanctuary.

Efforts to protect the Channel Islands began in 1938, when President Franklin D. Roosevelt designated Santa Barbara and Anacapa Islands as Channel Islands National Monument. Recognizing the cultural and natural significance of the islands, in 1980 the National Park Service designated San Miguel, Santa Rosa, Santa Cruz, Anacapa, and Santa Barbara islands, plus the waters within one nautical mile of each island, as the nation's 40th national park. NOAA designated the ocean waters surrounding the five islands as Channel Islands National Marine Sanctuary later that year. In 2003 the state of California created a network of marine protected areas within the nearshore waters of the Channel Islands National Marine Sanctuary, and NOAA expanded the network into the sanctuary's deeper waters in 2006 and 2007. Today the entire network consists of 11 marine reserves where all take and harvest is prohibited, and two marine conservation areas that allow limited take of lobster and pelagic fish.

Gulf of Mexico

While cold water dominates the coastal oceans of the West Coast, warm water bathes the Gulf of Mexico and the southern Atlantic Coast of the United States. The Gulf of Mexico is the origin of the mighty Gulf Stream, a warm current so powerful it reaches all the way across the Atlantic to Europe, making it possible for palm trees to grow on England's southern shores.

The Gulf of Mexico is ecologically diverse and highly productive. Its waters house otherworldly species: tubeworms that evolved to thrive in the gaseous environment, long-legged spider crabs hunting for dinner, mussels bellying up to the methane bar for energy. Sharks, skates, and rays glide in the water column. And dolphins roam the waters, surfing in the wake of ships.

The only national marine sanctuary in the Gulf of Mexico lies 70 to 115 miles off the Texas and Louisiana coasts. Snapper and grouper fishermen visiting the area in the early 1900s saw brightly colored sponges, plants, and other marine life below their boats, and the area became known as "flower garden banks." The name stuck. On a 1960 expedition, scientists confirmed that the banks had not only living coral but also massive, thriving coral reefs. About 30 years later, Flower Garden Banks National Marine Sanctuary was designated. The sanctuary consists of three banks connected by currents, some of which carry coral spawn and fish larvae, algal spores, and other sea life east toward Florida. Because they are far offshore and less impacted by humans than most coral reefs, the corals on and near the Flower Garden Banks are among the healthiest in US water. This also makes them a living laboratory for coral research. As of the writing of this book, NOAA is proposing to expand Flower Garden Banks from 56 square miles to 206 square miles covering 17 banks.

This little piece of the Caribbean in the northern Gulf of Mexico is a prime diving destination for those willing to make the lengthy journey offshore. Manta rays, whale sharks, sea turtles, and stingrays all provide exciting encounters, as do the colorful reef fish and invertebrates that call these massive coral reefs home. Every August divers compete for the few spots available to see the annual mass coral spawning that takes place for only seven to ten nights after the full moon. This spectacular underwater snowstorm is the corals' way of ensuring genetic diversity and dispersal.

At the eastern end of the Gulf lies Florida Keys National Marine Sanctuary. Beginning south of Miami, it spans 220 miles, linked by 42 bridges over aquamarine water, to end in Key West. The sanctuary protects a mosaic of marine communities, including a nearly continuous line of offshore reefs washed by the warm, clear waters of the Gulf Stream—the only bank and barrier reef that lie along the US coast. Closer to shore, patch reefs are interspersed with seagrass meadows and coastal forests of mangrove trees, the only tropical trees that can grow right in saltwater. Groupers, snappers, jacks, and sharks

A diver meets the local amberjacks at Flower Garden Banks National Marine Sanctuary. The sanctuary includes the northernmost coral reefs in the continental United States. The reefs and banks support critical deepwater habitats, including brilliant reef-building corals and sponge assemblages atop underwater mountains, and a beautiful bounty of wildlife.

dominate the offshore reefs. Young fish hide among knobby-rooted mangroves where herons, egrets, and other wading birds hunt for bite-sized morsels. Manatees and green sea turtles graze in seagrass meadows, which also shelter spiny lobsters, shrimp, starfish, sea anemones, and sponges. In all, the sanctuary protects more than 6,000 species of marine life and an estimated 400 underwater historic sites, including 14 listed on the National Register of Historic Places.

At the southwestern tip of the sanctuary, 70 miles beyond Key West, lies Tortugas Ecological Reserve. The healthy, deep coral reefs here are the crown jewels of the Florida reefs. In the late 1990s, a group of Keys residents worked with NOAA to establish the 200-square-mile reserve, the largest no-take marine reserve in the continental United States. The reserve does not permit fishing or wildlife harvesting of any kind, and strictly controls anchoring to protect the reefs from physical damage. These protective measures are working. More and larger fish are found inside the reserve and just outside its boundary than in other parts of the Florida Keys. And mutton snappers, once threatened by overfishing, now gather in large groups in the reserve to mate.

East Coast

As the Gulf Stream leaves the Keys and moves up the Atlantic Coast, the ocean's color changes from clear, tropical aquamarine to darker greens and grays. The Atlantic Coast's geology is very different than the Pacific's, and the differences are evident from the coast to the ocean floor. Here, the nearest tectonic plate boundary is far away in mid-ocean, and the shoreline gives way to a gently sloping continental shelf. During the last Ice Age, when much more of the planet's water was confined in glaciers and ice caps, this shelf was dry land stretching up to 150 miles beyond today's shorelines. Now barrier islands face the sea. Behind them rivers, bays, and salt marshes shelter menhaden, red drum, crabs, shrimp, and other species. Sea stars, sea anemones, barnacles, whelks, and mussels cling to rocks, and shorebirds feed between tides on the shorelines. On the seafloor, low-relief ledges break up shallow, sandy basins. These ledges, closer to the warmth and light of the sea surface, are encrusted with crabs, shrimp, and mollusks such as clams, squids, slugs, and snails.

Many of the Atlantic sanctuaries lie miles offshore. When sea levels were lower, Gray's Reef, off Georgia, and Stellwagen Bank, off Massachusetts, were terrestrial hills, bluffs, or plateaus. Now submerged rock formations crowned with living corals lie scattered across the southern Atlantic continental shelf. At Gray's Reef National Marine Sanctuary, located 19 miles off the coast in about 70 feet of water, you find one of these formations. At 22 square miles, it is one of the smallest sites in the system. Its rocky ridges are riddled with caves and crevices where sea squirts, barnacles, snails, corals, sea stars, lobsters, and other invertebrate animals shelter. These animals form a dense carpet of living creatures that in places completely hides the rock, giving the

Dive or fish in Gray's Reef National Marine Sanctuary, and you're likely to come across many black sea bass. In Gray's Reef, black sea bass play an important role as predators, keeping populations of crabs, shrimp, and small fish in check. As protogynous hermaphrodites, black sea bass will switch sexes as they mature, generally starting out as female and changing to male.

habitat of Gray's Reef its name: "live bottom." Groupers, snappers, and other fish gather on the reef to feed. Loggerhead sea turtles, a dozen kinds of sharks, and marine mammals including the highly endangered right whale are among the large sea creatures drawn to this rich feeding zone.

Nearly one-third of Gray's Reef is set aside as a research area, the largest percentage reserved for science in any sanctuary. In research reserves, managers control activities to minimize their effects on marine life and allow scientists to get a clearer view of the effects of phenomena such as climate change, and of human activities such as fishing. At Gray's Reef in 2019, scientists were tracking the numbers of fish, their diversity, and their distribution around different types of undersea formations, some of which are packed with prey fish and their predators. They were studying tiny microbes and large macroalgae, or seaweed, and on the lookout for threatened sea turtles, as well as unwanted species such as lionfish, a voracious invader from Asian waters.

Gray's Reef is not a typical reef. Rather than the hard coral characteristic of tropical reefs, its foundation is formed from the consolidation of sediment like shell fragments, sand, and mud. These sediments were borne on ocean currents and deposited as a blanket of loose grains starting 6 million years ago.

Approximately 2 million years ago, briny seawater containing naturally occurring calcium-and-carbon compounds from the shells of marine animals provided the glue that stuck these sediments together, creating cemented sandstones that make spaces for an abundance of marine life. Scientists and researchers have also discovered countless Pleistocene fossils, such as mammoths, from before the Ice Age, when terrestrial creatures roamed the dry land that now is Gray's Reef. Relics of that ancient time lie far below the ocean surface, waiting to be found.

North of Gray's Reef, and more than 200 feet underwater, lies the Civil War ironclad ship USS *Monitor*. This famous warship is the focus of Monitor National Marine Sanctuary, off the coast of Cape Hatteras, North Carolina. Built quickly in response to news that the Confederacy was about to launch an iron-armored warship, *Monitor* emerged from a New York shipyard in January 1862 and voyaged to Hampton Roads, Virginia. There, in March 1862, it fought the Confederate ironclad CSS *Virginia* to a draw that both sides claimed as a victory.

Monitor became a symbol of American ingenuity. But at the end of 1862, in a storm off Cape Hatteras, it disappeared. Sixteen crew members were lost along with it. The wreck was discovered in 1973, after decades of fruitless searches, by a team of scientists led by Duke University's John G. Newton. It lay 16 miles offshore, beyond the protection of state and federal laws—save one, the National Marine Sanctuaries Act. So in 1975 President Gerald Ford designated the remains of *Monitor* and a surrounding column of water one mile in diameter as the nation's first national marine sanctuary.

Near Monitor National Marine Sanctuary is an area known as the "Graveyard of the Atlantic," where thousands of vessels and countless mariners were lost. Researchers are documenting hundreds of vessels that sank in this area during World War II's Battle of the Atlantic, one of the longest and most decisive campaigns of the war. The war came home to the continental United States on the Outer Banks, and the shipwrecks there are an important piece of US maritime history.

Mallows Bay–Potomac River National Marine Sanctuary, just 40 miles south of Washington, DC, is the newest sanctuary, designated in 2019, and the only sanctuary in the Chesapeake Bay watershed. It captures roughly 12,000 years of human habitation on the banks of the Potomac River, from ancient Native American artifacts and sites through Revolutionary and Civil War battle sites, remnants of Potomac River steam-powered ferries, and historic commercial fishing operations. But the sanctuary is most renowned for the "Ghost Fleet"—partially submerged remains of more than 100 World War I wooden steamships constructed to carry food and troops supplies for the Allies. Built at more than 40 shipyards in 17 states between 1917 and 1919 for the US Emergency Fleet, the steamships were part of a massive national wartime mobilization

Over 40 species of seabirds regularly come to Stellwagen Bank's waters to feed on the abundant fish, plankton, and other marine invertebrates. This Wilson's storm petrel is a long-distance migrator; it breeds in Antarctica and ranges to temperate and subarctic areas of the Northern Hemisphere.

On the Wings—and Breath—of a Bird

Several thousand seabirds wheel around Stellwagen Bank National Marine Sanctuary's research vessel, the *Auk*, while humpback whales break the sea's surface, their open mouths streaming water and fish. Peter Hong doesn't even glance at the ocean. He is busy fitting a small mask over the bill of a great shearwater. The mask, which is connected to a small vacuum chamber, will collect a few of the bird's exhaled breaths for analysis.

Great shearwaters are medium-sized seabirds with four-foot wingspans. They are world travelers, spending November through April in the Southern Hemisphere, where they nest on the remote Tristan da Cunha islands in the South Atlantic. In April, adults and young birds make their way north. They appear at Stellwagen Bank National Marine Sanctuary and in the surrounding Gulf of Maine in June.

Hong and others on the sanctuary research team are waiting for them. The scientists use cut-up fish and squid to lure the birds close to a small boat, net them, and take them to the *Auk*, where they carefully collect blood and feather samples from each bird. The samples reveal what the birds have been eating in recent weeks and months. But collecting and testing the birds' exhaled gases is a new technique, one that can show what the birds have just eaten. Scientists also attach small satellite tags to a few birds' backs to track their movements for months.

Researchers want to understand how great shearwaters are foraging and feeding here and elsewhere on their 6,500-mile migration. The research can help sketch a picture of environmental disturbances brought about by global climate change, overfishing, and other factors. As top predators, seabirds are good indicators of changes in food webs and the overall health of the marine ecosystem.

to build 1,000 ships in 18 months. The war ended before the ships saw action, and the government eventually sold the ships to a salvage company that moved them to Mallows Bay.

From a bird's-eye view, the sanctuary is dotted with small, lens-shaped islands. Some are completely covered with vegetation, and on others, the wooden skeletons of the Ghost Fleet peek through the water. The Ghost Fleet remains provide habitats for many of the Chesapeake Bay's native species and offer opportunities to see more than 100 species of native or migratory birds: great blue heron, bald eagle, golden ibis, and canvasback duck. Kayakers paddling among the shipwrecks glimpse sharp-clawed, yellow-eyed osprey stationed atop the weathered bowsprit of a ship that has not sailed in more than 100 years. Box turtles peep their heads out of the cool water, occasionally climbing onto a fallen tree branch to let the sun warm their shells.

Mallows Bay–Potomac River National Marine Sanctuary is also the traditional homeland and cultural landscape of the Piscataway Indian Nation, Piscataway Conoy Confederacy and Sub-Tribes of Maryland, and the Patawomeck Indian Tribe. Although no archaeological sites have been identified in waters protected by the sanctuary, it is likely that Nussamek, one of the villages visited by the English explorer Captain John Smith during the summer of 1608, is in this area.

Though Smith is best known for his exploration of the Chesapeake Bay and its Tidewater region, and his role in founding Virginia's Jamestown colony in 1607, a later voyage took him to New England, where he made important discoveries. In 1616 Smith published a map of New England showing a single ship—the mapmaker's sign of good fishing grounds—placed over what is now Stellwagen Bank National Marine Sanctuary. Early European fishing crews were drawn to the bank by the huge congregations of seabirds and marine mammals feeding there. "The abundance of Sea-Fish are almost beyond believing," wrote Reverend Francis Higgins in 1629, in *New Englands Plantation or, A Short and True Description of the Commodities and Discommodities of that Countrey*. "I saw great store of Whales . . . and such abundance of mackerels that it would astonish one to behold." In 1635 a reprint of Smith's map reinforced the region's reputation for bountiful seas by placing a pyramid of cod heads under the ship.

The exact number of banks found in the North Atlantic is unknown, but Stellwagen Bank is one of the most ecologically and economically important. The bank, a submerged sand and gravel plateau six miles north of the tip of Cape Cod, was once dry land. At Stellwagen Bank, deep, nutrient-rich waters mix with warmer surface waters, attracting more than a dozen marine mammal species. Humpback whales gather here in droves each summer and fall to feed on a small fish called the sand lance. More than three dozen seabird species, from high-soaring fulmars and gannets to swift underwater-swimming alcids, fly over

Divers explore the steam engine on the SS *Florida*. During a dense fog in May 1897, the steamer *Florida* collided with another steamer, the *George W. Roby*, and sank—nearly cut in half by the collision. The *Florida* rests upright and under 200 feet of water in Lake Huron off Presque Isle, Michigan, in Thunder Bay National Marine Sanctuary.

sanctuary waters and feed in its waters. On the bank, sand dollars, clams, shrimp, worms, and tunicates shoulder over one another in an ever-shifting kaleidoscope of life. Strange-looking flatfish such as halibut and flounder inhabit the bottom, while sleek, powerful bluefin tuna circle around and above the plateau.

Great Lakes

Wild, tidal, and as storm-tossed as the open ocean, the Great Lakes are nicknamed the "Third Coast" of the United States. These five enormous freshwater bodies, connected to the Atlantic Ocean through the St. Lawrence Seaway, hold an estimated one-fifth of the fresh water on Earth, an unfathomable six quadrillion gallons. French explorers in 1615 called Lake Huron *la mer douce*, which means "the sweet sea," because of its fresh water and vastness. There, located along northwestern Lake Huron, you find Thunder Bay National Marine Sanctuary.

For more than 12,000 years, people have traveled on the Great Lakes in all sorts of crafts, from Native American dugout canoes to wooden sailing vessels and steel-hulled freighters. The past 150 years have been particularly explosive, transforming this inland sea into one of the world's busiest waterways. Yet with extraordinary growth comes adversity. More than 200 pioneer steamboats, majestic schooners, and huge steel freighters wrecked near Thunder Bay alone. The sanctuary was created primarily to protect a trove of archaeological relics and shipwrecks within 4,300 square miles of waters aptly known as "Shipwreck Alley."

A dive into the graveyard of shipwrecks is an encounter with history—every ship tells a new story. The sunken vessels supplied Native peoples and early settlers with the goods upon which they built their societies. Later they carried the iron ore, lumber, and coal that built a modern nation. The sanctuary's shipwrecks capture dramatic moments from the centuries that transformed America from a lightly inhabited wilderness to an industrial powerhouse. Taken together, they illuminate an era of enormous growth and remind us of the risks taken and tragedies endured. Lake Huron's cold, fresh water ensures that Thunder Bay's shipwrecks are among the best preserved in the world. With masts still standing, deck hardware in place, and the crews' personal possessions often surviving, sites located in deeper waters are true time capsules. Other shipwrecks lie well preserved but broken up in shallower waters.

Underneath the waters of national marine sanctuaries are amazing wonders, from shipwrecks that tell the story of the past, to miraculous animals that dwell in hidden seascapes. The National Marine Sanctuary System protects the wildness and wonder of our ocean and Great Lakes through exploration and science, resource conservation, education, and stewardship. It is a unique legacy given to future generations. ○

IMAGES OF WONDER

A Photographic Portfolio of the National Marine Sanctuaries and Monuments

THE PACIFIC

NATIONAL MARINE SANCTUARY OF AMERICAN SAMOA

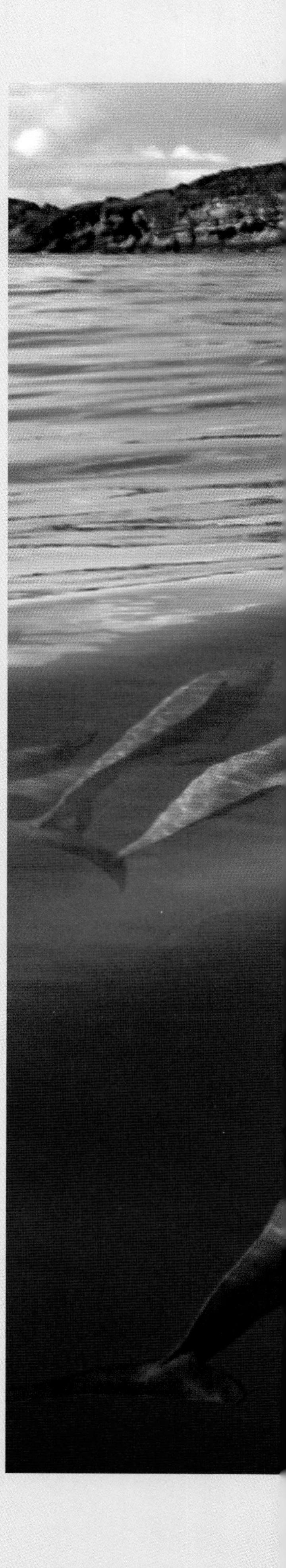

ABOVE

Located in the South Pacific Ocean, Rose Atoll Marine National Monument covers 8,571,633 acres and encompasses the Rose Atoll National Wildlife Refuge. This photo, showing coral in the shallow waters of Rose Atoll, appears in the film *Hidden Pacific.*

RIGHT

These spinner dolphins are heading toward the small volcanic island of Aunuʻu, one of several marine areas added to the 13,581-square-mile National Marine Sanctuary of American Samoa when the sanctuary expanded in 2012.

The reefs of National Marine Sanctuary of American Samoa, like this pink and green reef surrounding Swains Island, provide shelter and habitat for tropical fish of all shapes, sizes, and colors, as well as crustaceans, squid, sharks, and sea turtles.

A red-spotted guard crab protects its home and food source, cauliflower coral, in National Marine Sanctuary of American Samoa. Guard crabs like these help protect corals; in exchange for shelter and food, the crabs ward off predators to the corals, such as snails and sea stars.

THE PACIFIC

HAWAIIAN ISLANDS HUMPBACK WHALE NATIONAL MARINE SANCTUARY

ABOVE

Humpback whales migrate 3,000 miles from the Gulf of Alaska to the Hawaiian Islands, where they spend the months of December through May. The whales are readily visible from beaches, and their acrobatics delight and amaze residents and visitors.

RIGHT

Hawaiian Islands Humpback Whale National Marine Sanctuary was designated to protect humpback whales and their habitat. Although the North Pacific humpback population is increasing, humpbacks remain endangered.

At cleaning stations such as this one, which are used by fish and sea turtles, the cleaner fish remove and eat the parasites from the larger animals' skin, sometimes even swimming into their mouths and gills.

THE PACIFIC

PAPAHĀNAUMOKUĀKEA MARINE NATIONAL MONUMENT

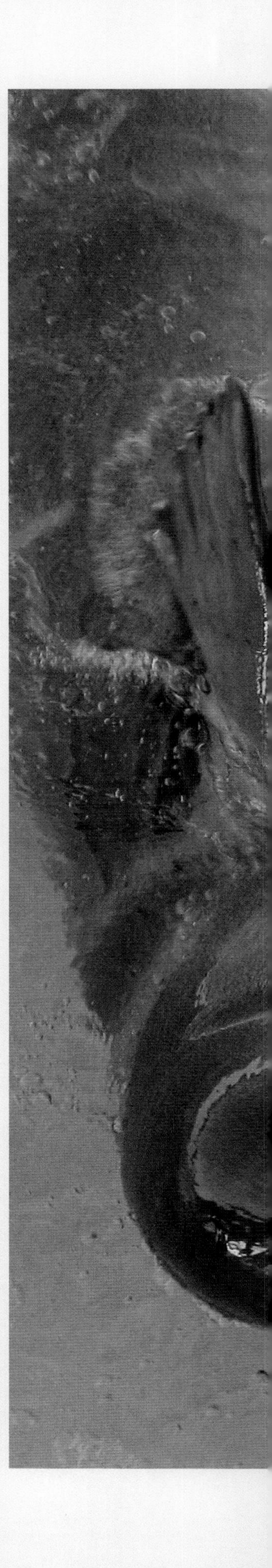

ABOVE

Deep-sea corals and sponges provide habitat and refuge for many other animals living on or near the seafloor. Here a sponge covered with hundreds to thousands of tiny anemones also is home to several brittlestars (pink), crinoids or "sea lilies" (yellow), and a basket star (brown).

RIGHT

Papahānaumokuākea Marine National Monument provides sanctuary to at least 23 endangered species, including the Hawaiian monk seal, which is found only in Hawai'i. This population of seals represents one of only two monk seal populations remaining anywhere in the world. The population of Mediterranean monk seals is perilously low; the monk seals of the Caribbean are extinct.

Pearl and Hermes Atoll has the highest number of fish species in the Northwestern Hawaiian Islands. The milletseed butterflyfish shown here are endemic to Papahānaumokuākea Marine National Monument.

THE WEST COAST

OLYMPIC COAST NATIONAL MARINE SANCTUARY

ABOVE

Ruby Beach is one of the most visited areas of Olympic National Park. The park overlooks Olympic Coast National Marine Sanctuary, which provides a safe haven for marine species.

RIGHT

Twenty-nine species of marine mammals reside in, or migrate through, Olympic Coast National Marine Sanctuary, including the majestic orca. Orcas, more commonly known as killer whales, are actually the largest members of the dolphin family. One of the most widely distributed mammals on Earth, the orca is found in all the ocean basins of the world.

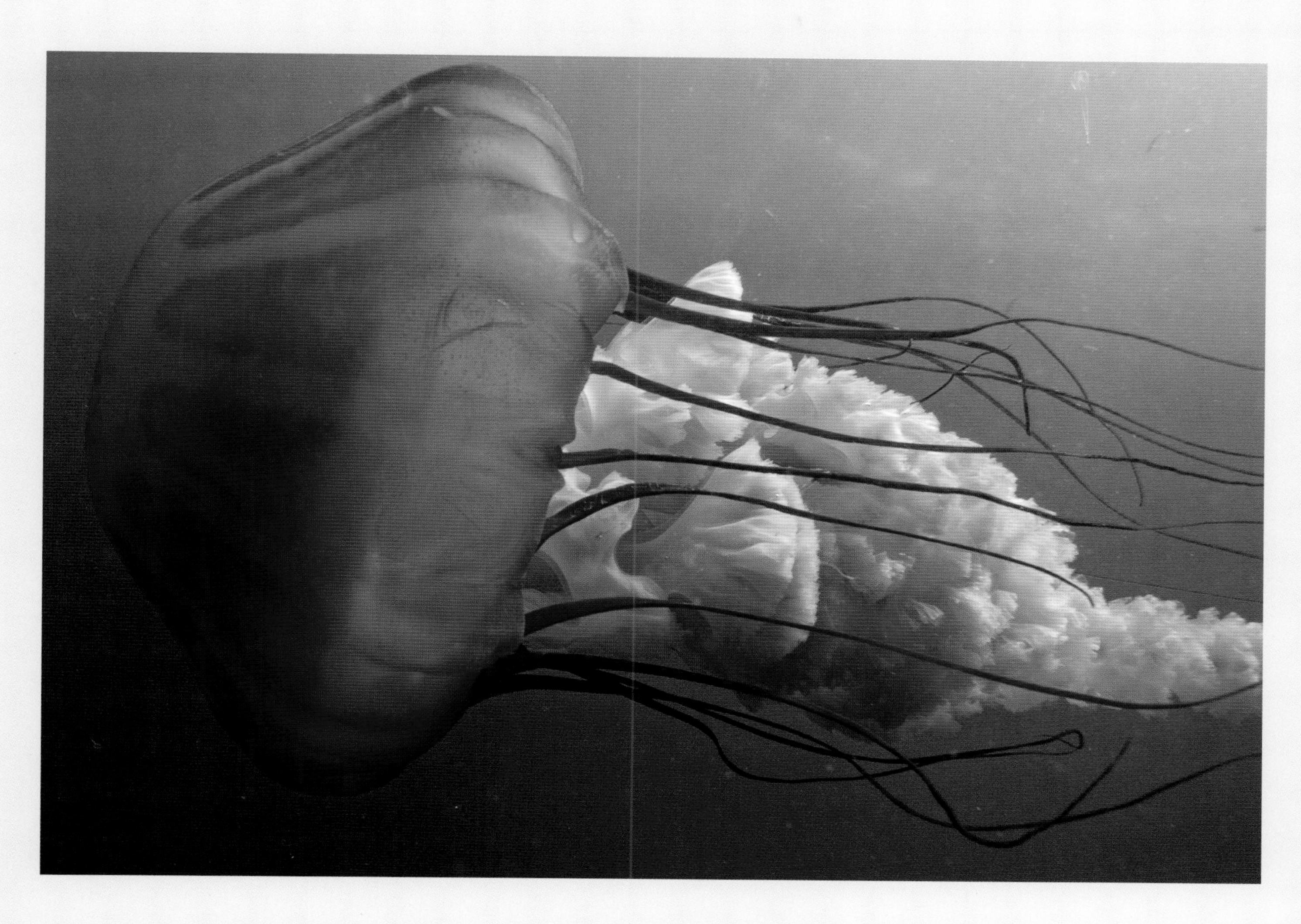

ABOVE

This brightly colored Pacific sea nettle swims using propulsion: it squeezes its bell, pushing water behind it. Sea nettles sense light, which they use to travel from dark, deep waters to shallow, sunlit waters each day. Sea nettles can be up to 30 inches wide and 16 feet long.

LEFT

The tide pools of Olympic Coast National Marine Sanctuary are perfect places for kids and adults alike to explore marine life, including these brightly colored sea urchins.

The Olympic Coast is home to the giant Pacific octopus, the largest known species of octopus. These highly intelligent creatures hunt at night for shrimp, clams, lobsters, and fish.

THE WEST COAST

CORDELL BANK NATIONAL MARINE SANCTUARY

ABOVE

The glassy, rolling surface of Cordell Bank National Marine Sanctuary reflects a common murre, one of the more than 50 seabird species that flock to this patch of Pacific Ocean 20 miles off Point Reyes, California.

RIGHT

In 1978, Robert Schmieder described his first glimpse of Cordell Bank as an overwhelming visual experience. It is the same today, as fish blanket the lush undersea gardens of colorful sponges, sea anemones, sea stars, sea cucumbers, snails, and crabs.

Bright pink strawberry anemones and barnacles cling to the rocks in Cordell Bank National Marine Sanctuary. Cordell Bank sits at the edge of the continental shelf and rises abruptly from the soft sediments to within 115 feet of the ocean surface.

THE WEST COAST

GREATER FARALLONES NATIONAL MARINE SANCTUARY

ABOVE

Greater Farallones National Marine Sanctuary, in collaboration with federal, state, and local partners, is restoring Bolinas Lagoon to protect and conserve this ecosystem. The lagoon, which is part of the sanctuary, is a wetland of international significance.

RIGHT

Sandpipers such as these sunset-lit birds on Manchester Beach probe into mud or shallow water with their bills to feed on insects, crustaceans, mollusks, and worms.

ABOVE

The mola mola is the biggest bony fish in the sea. This one was spotted in Greater Farallones National Marine Sanctuary. Mola molas spend time basking on their sides near the surface, with their pectoral fins flapping in the air.

LEFT

Steller sea lions live in several national marine sanctuaries, including in Greater Farallones National Marine Sanctuary, which protects one of the southernmost populations in the US. Steller sea lions are vocal mammals which bark, growl, and grunt. Their noisy voices fill the air.

THE WEST COAST

MONTEREY BAY NATIONAL MARINE SANCTUARY

ABOVE

Stretching from Marin to Cambria, California, Monterey Bay National Marine Sanctuary encompasses a shoreline length of 276 miles. The Bixby Bridge portion of Big Sur's coastline is in the southern portion of the sanctuary.

LEFT

During the spring and summer, northwest winds push surface water offshore, replacing it with cold water from greater depths. These waters, rich in nutrients, fuel the growth of plankton. Whales, dolphins, and seabirds concentrate in Monterey Bay National Marine Sanctuary to feed on an abundance of krill, fish, and squid supported by these intense plankton blooms.

ABOVE

Cabezon, which means “large head” in Spanish, is an appropriate name for this fish. Often found in the kelp around rocky reefs, cabezon will eat crustaceans, fish, and mollusks that fit in its mouth. This handsome cabezon was spotted at Point Lobos State Marine Preserve, located in Monterey Bay.

RIGHT

This mass of fish larvae, known as ichthyoplankton, was found in Monterey Bay. The eggs drift in the ocean, flowing along with water currents. Monitoring fish eggs and larvae can provide scientists with clues on the health of an ecosystem.

Sea otters, such as this one feasting on a gigantic crab in Monterey Bay, eat about 20 to 30 percent of their body weight each day. They play a critical role in maintaining healthy marine ecosystems by eating sea urchins and other invertebrates that graze on giant kelp.

THE WEST COAST

CHANNEL ISLANDS NATIONAL MARINE SANCTUARY

ABOVE

In the distance, the sand dunes at Bechers Bay on Santa Rosa Island fade into Channel Islands National Marine Sanctuary.

RIGHT

A moray eel emerges from a rock crevice. While most fish breathe through gills, the moray eel must constantly open and close its mouth to breathe, making it look as if it is gasping for air. Its open mouth also showcases the moray's fang-like teeth, which have led to exaggerated fears of a creature that is shy around humans.

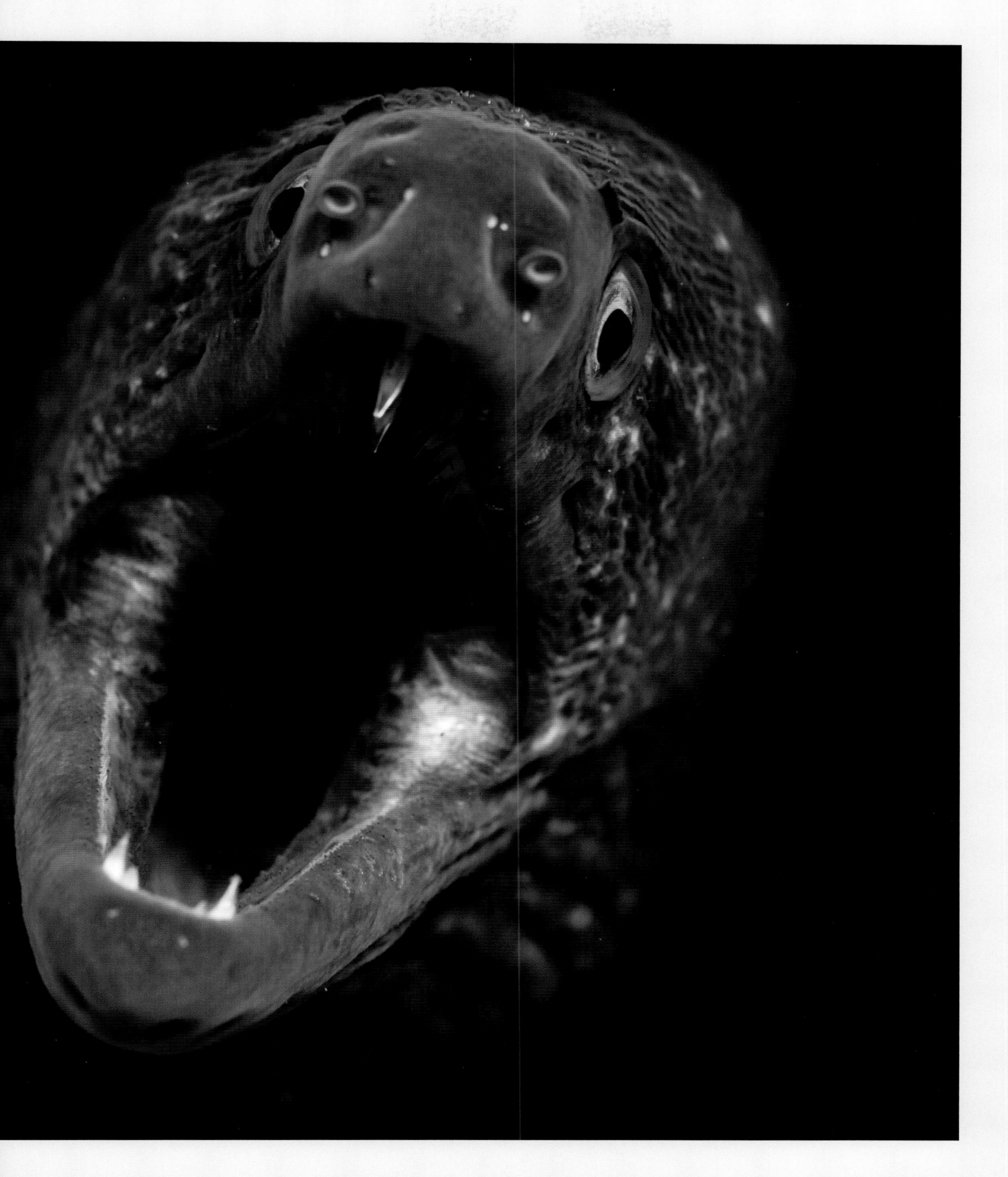

Prized by fishers and divers, the California spiny lobster lives in the kelp forests and surfgrass beds of Channel Islands National Marine Sanctuary. The lobsters help maintain the diversity of intertidal and subtidal communities.

THE GULF OF MEXICO

FLOWER GARDEN BANKS NATIONAL MARINE SANCTUARY

ABOVE

Christmas tree worms are found throughout Flower Garden Banks National Marine Sanctuary. The two "trees" are the gills of the worm, whose body is embedded in the coral.

LEFT

Researchers can identify manta rays by their distinctive underside markings. More than 60 individuals have been added to the catalog of mantas that have been spotted at this "hot spot" for marine life in the Gulf of Mexico.

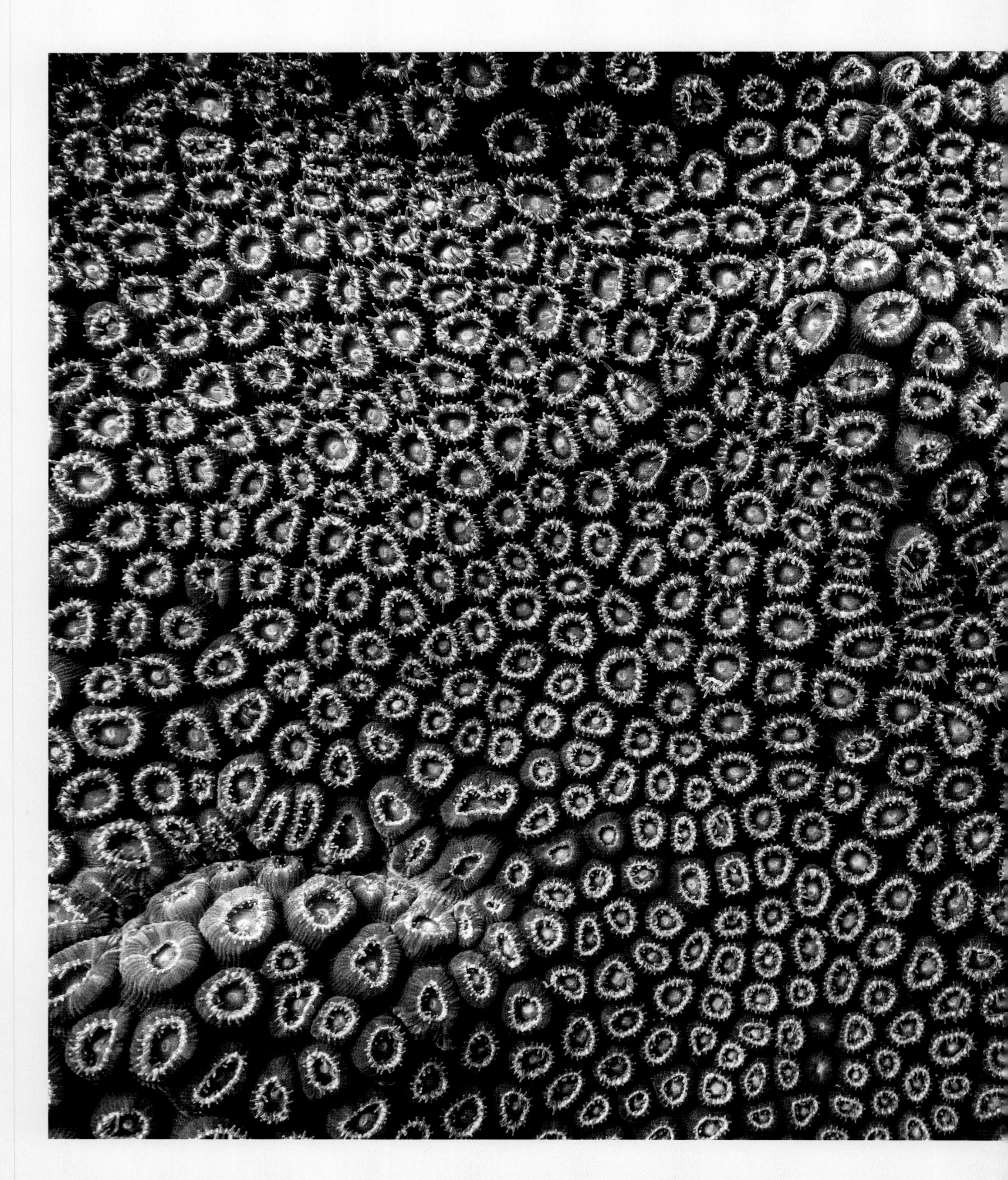

ABOVE

Resembling a burst of celebratory confetti, corals like this one spawn within Flower Garden Banks National Marine Sanctuary each year, releasing hundreds of gametes into the water. The warm, sunlit waters of this Gulf of Mexico sanctuary make it a comfy home for hard corals like these, as well as hundreds of other marine species.

LEFT

The corals on and near the Flower Garden Banks are among the healthiest in US waters, which makes them a living laboratory for coral research. This lobed star coral is a threatened species.

THE GULF OF MEXICO

FLORIDA KEYS NATIONAL MARINE SANCTUARY

ABOVE

Carysfort Lighthouse, which marks a shallow reef off Key Largo in Florida Keys National Marine Sanctuary, is visible in the distance, surrounded by white clouds and blue skies.

RIGHT

Manatees migrate to Florida Keys National Marine Sanctuary to winter in its warm, shallow waters.

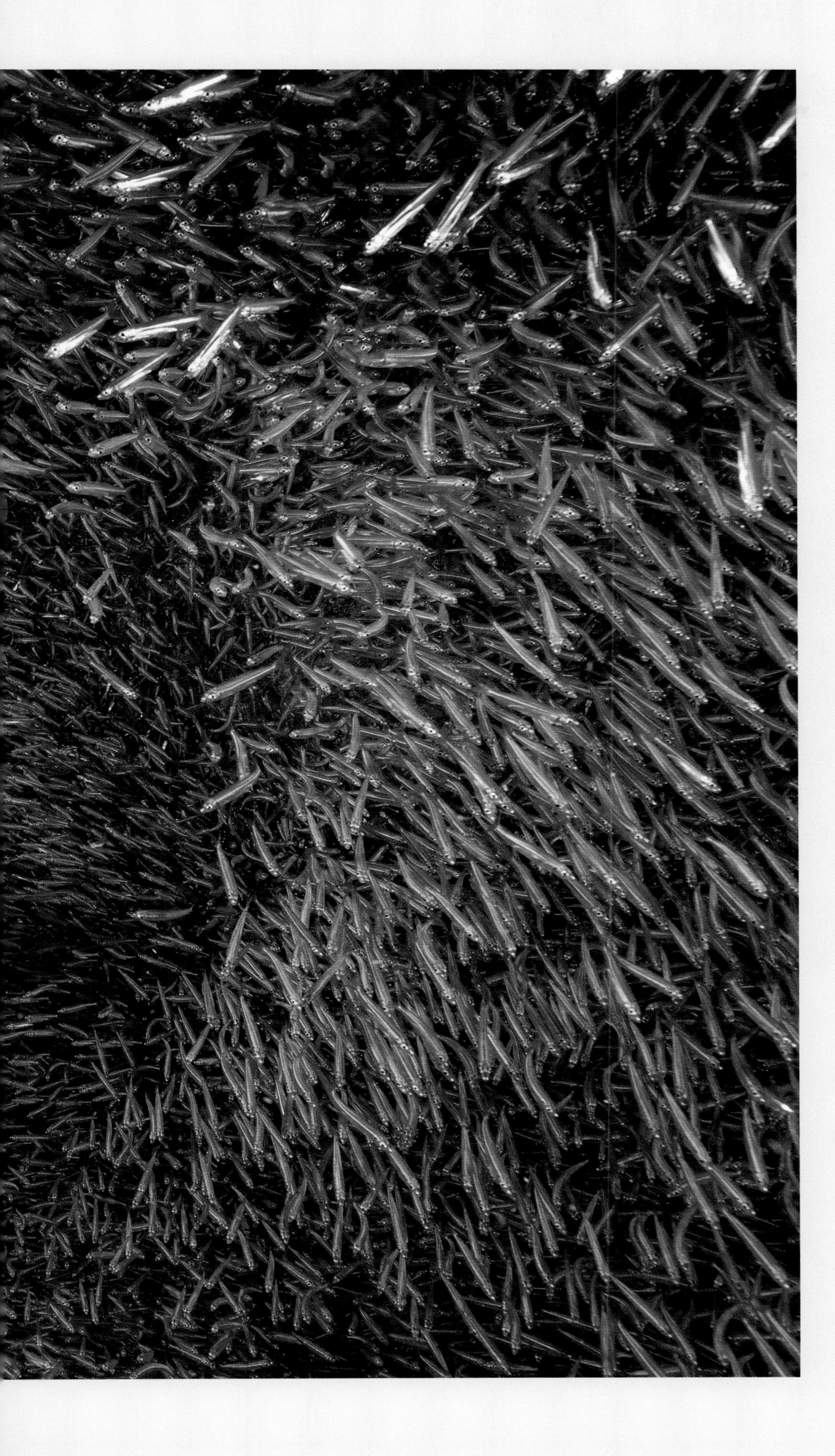

A barracuda swims through a mass of silvery fish. With a long, thin body and a set of fang-like teeth, these fish often evoke fear, but to experienced divers, barracuda are more famous for their curious nature and for being camera-friendly.

THE EAST COAST

GRAY'S REEF NATIONAL MARINE SANCTUARY

ABOVE

Curtains of soft corals brighten Gray's Reef, which is currently the only protected natural reef area on the continental shelf, off the Georgia coast.

RIGHT

A small school of fish and even a shrimp hitch a ride with this jellyfish traveling through the waters of Gray's Reef. Some juvenile fish can live amid a jelly's tentacles without being harmed.

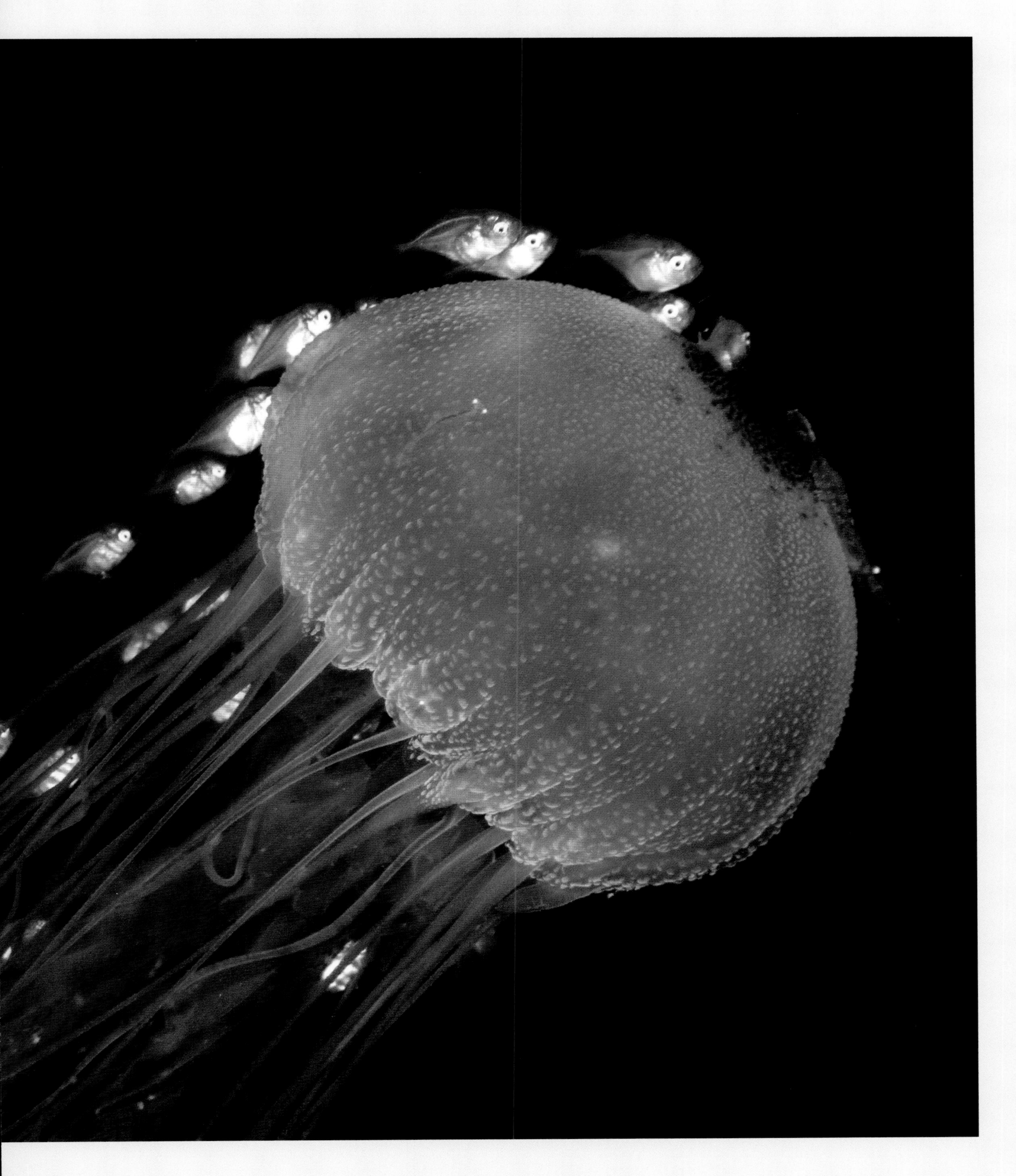

Two blue crabs make themselves known as their striking red claws stand out in a sea of gray. Their scientific name, *Callinectes sapidus*, means "savory, beautiful swimmer," and they can range from as far north as Nova Scotia as far south as Uruguay.

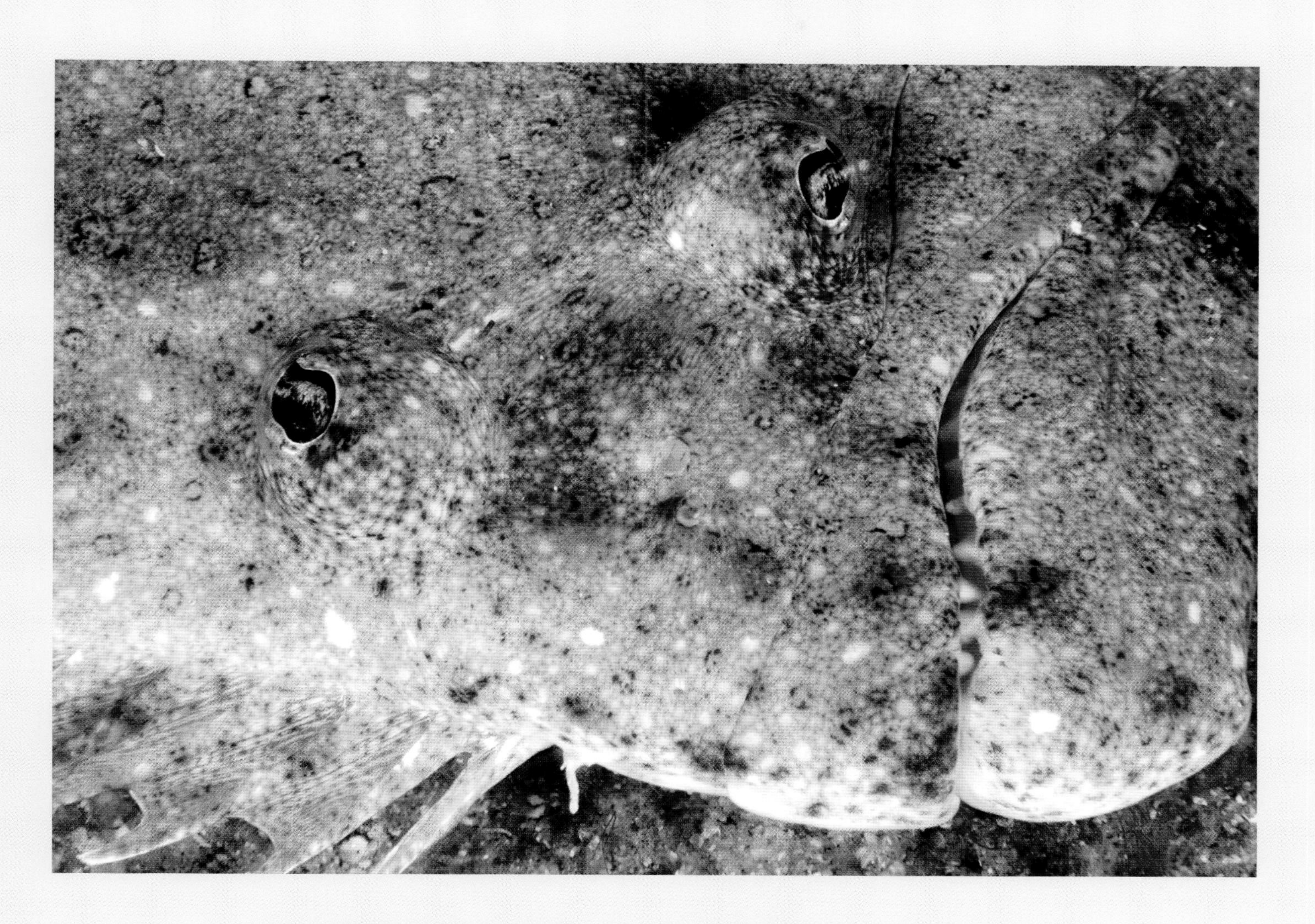

A master of disguise, a dusky flounder nestles in sand indistinguishable from its own skin—only its protruding eyes give it away. Living on the bottom of the ocean, the dusky flounder patiently scans the waters above until it is ready to ambush small fish and invertebrates passing by.

THE EAST COAST

MONITOR NATIONAL MARINE SANCTUARY

ABOVE

Buried more than 200 feet underwater lies the Civil War ironclad ship USS *Monitor*. Built quickly in response to news that the Confederacy was about to launch an iron-armored warship, *Monitor* emerged from a New York shipyard in January 1862 and voyaged to Virginia. There, in March 1862, it fought the Confederate CSS *Virginia* to a draw. *Monitor* sank in a storm less than a year later.

LEFT

Schools of fish swirl around the crow's nest of a tugboat wreck near Cape Lookout, North Carolina. The wreck is in an area under consideration for protection as part of an expansion of Monitor National Marine Sanctuary.

Among the shipwrecks in the area proposed for expansion around Monitor National Marine Sanctuary are many from World War II's Battle of the Atlantic, including this wreck of a U-701. Sunk on July 7, 1942, by a US Army Air Forces Hudson, this U-boat sits in approximately 110 feet of water off of Cape Hatteras, North Carolina.

THE EAST COAST

MALLOWS BAY–POTOMAC RIVER NATIONAL MARINE SANCTUARY

ABOVE

Elementary school students gather around a fish tank to learn about different fish species and other marine wildlife found in the region during a field trip to Mallows Bay–Potomac River National Marine Sanctuary.

LEFT

A World War I "Ghost Fleet" of more than 100 wooden steamships lies stranded along an 18-mile section of the Potomac River at Mallows Bay–Potomac River National Marine Sanctuary. The war ended before the fleet was needed for service, and many of the ships were scuttled.

Kayakers explore Mallows Bay–Potomac River National Marine Sanctuary, which protects shipwrecks dating from the Revolutionary War to the present day.

Designated in 2019, Mallows Bay–Potomac River National Marine Sanctuary is the only sanctuary in the Chesapeake Bay watershed. Nature is slowly reclaiming the wooden steamships within the river and creating new habitat for fish, wildlife, and birds.

THE EAST COAST

STELLWAGEN BANK NATIONAL MARINE SANCTUARY

ABOVE

The whale tagging boat at Stellwagen Bank National Marine Sanctuary gets a treat as this humpback whale breaches right by them. Data from tagged whales help in better understanding their movements and behavior underwater, which will in turn enable managers to better protect these marine mammals from threats like entanglement and vessel strikes.

LEFT

The lion's mane jellyfish uses its stinging tentacles to capture, pull in, and eat prey such as the juvenile haddock seen here. This is the largest known species of jellyfish; it can range in size from less than an inch to over eight feet.

Blue sharks are a migratory species, traveling across ocean basins. Commonly found in the waters of Stellwagen Bank National Marine Sanctuary, the blue shark's slim body, long fins, and coloring provide camouflage in the water.

THE GREAT LAKES

THUNDER BAY NATIONAL MARINE SANCTUARY

ABOVE

The Shipwreck Rowers team passes the Alpena Light on their early foggy morning workout in a Heritage-23 Mackinaw boat. Alpena Light is part of the Michigan Lighthouse Conservancy, which promotes the preservation of Michigan's lighthouses and lifesaving stations.

RIGHT

A 143-year-old shipwreck, the *E. B. Allen*, lies in the cold, fresh waters of Lake Huron. The water temperatures slow the deterioration of wood and metal, resulting in the remarkably well-preserved collection of wrecks protected by Thunder Bay National Marine Sanctuary.

SS *Portland*, a wooden schooner that wrecked in Thunder Bay in 1877, has gradually been torn apart by waves and storms over the years.

3

STORIES BENEATH THE WAVES

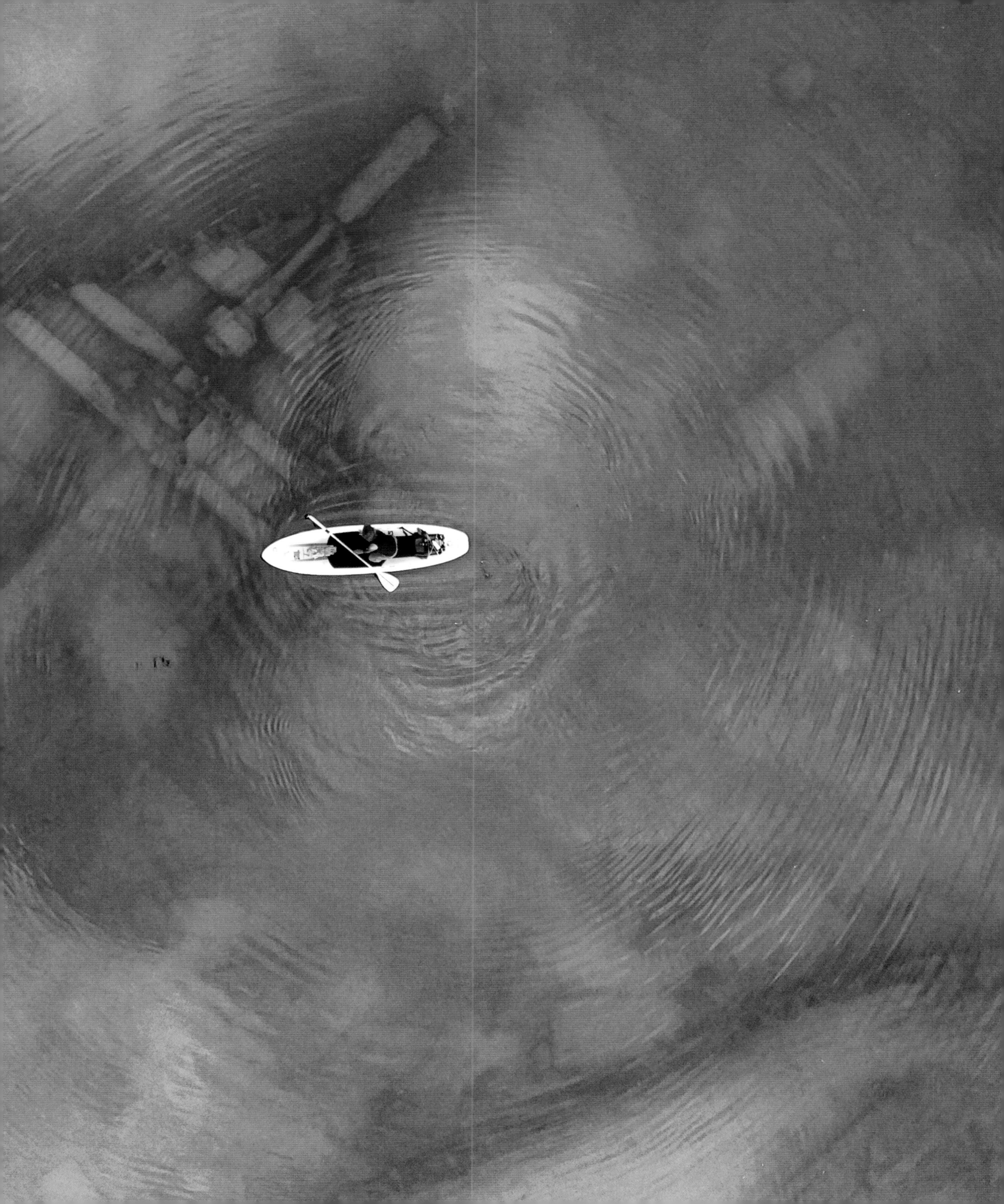

I could never stay long enough on the shore.
The tang of the untainted, fresh and free sea air
was like a cool, quieting thought.

HELEN KELLER

HUMANS ARE MARINERS, FISHERS, EXPLORERS, and harvesters, leaving traces of coastal habitation, exploitation, and wars scattered across the seascape. The waters of the sanctuary system preserve treasures equal to those in the world's greatest museums. They hold the submerged remains of ships, aircraft, and cultures, steeped in meaning because of our long history of reliance on the sea.

There's nothing quite like diving hundreds of feet below the water's surface to explore a shipwreck. Imagine descending through blue water that darkens as you go deeper, then catching a glimpse of a large structure covered in muted greens, reds, and browns. Fish hiding within the structure's crevices flit away as you get closer to the enormous shape. It is hard to see what's left of the vessel's original form because of the encrusting corals, barnacles, and sponges that creep up the sides and top. Maybe it's a Spanish galleon, part of a 16th-century treasure fleet that sailed from Havana into the howl of a hurricane and was grounded on the reef of what now is Florida Keys National Marine Sanctuary. Perhaps it's a 19th-century California coastal steamer, lost in a fog off the Golden Gate and now lying on the bottom of Greater Farallones National Marine Sanctuary. Or it could be the ancient hull of a Native American canoe, breached and beached thousands of years ago in a setting that once was a rocky shoreline in Channel Islands National Marine Sanctuary, but today lies below the surface, submerged by rising seas.

No matter what the wreckage turns out to be, the scientists of the sanctuary system's marine heritage program learn all they can about the vessel, its crew, the people that created it, and the catastrophe that destroyed it, while disturbing it as little as possible, and only rarely excavating pieces of the wreck. The same is true of the remnants of landward structures, like piers and moorings, fishing weirs, and chutes for lumber or ore. Think of each artifact as a page in the long story of humans' relationship with the sea. To remove it would be like tearing a page from a rare book: important parts of the meaning would be lost.

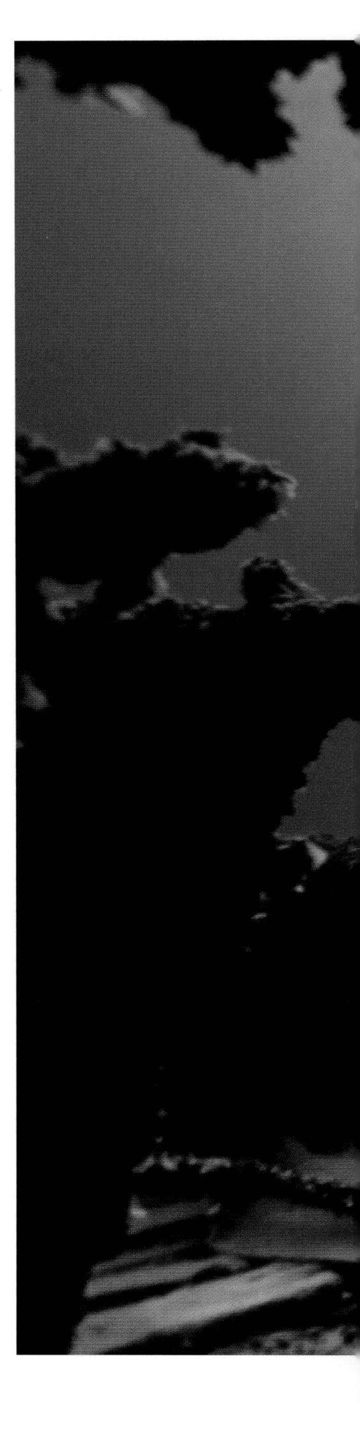

Located in northwest Lake Huron, Thunder Bay is adjacent to one of the most treacherous stretches of water within the Great Lakes system, earning it the nickname "Shipwreck Alley." Fire, ice, collisions, and storms have claimed more than 200 vessels in and around Thunder Bay. Here a diver inspects the wreck of *D. M. Wilson*, which sank in 1894.

PREVIOUS PAGES

At Forty-Mile Point, Presque Isle, Michigan, a stand-up paddle boarder crosses over the shipwreck *Joseph S. Fay* in Thunder Bay National Marine Sanctuary. The giant bulk freighter *Fay* hit the rocks at Forty-Mile Point during a gale in 1905 and sank. The freighter rests in 17 feet of crystal-clear water in Lake Huron.

Visible from San Francisco on the rare occasions when they aren't shrouded in fog, the Farallon Islands were sometimes called the "Devil's Teeth" by 19th-century sailors, who feared their jagged rocks and unpredictable currents.

For Hawai'i and the Pacific, WWII had a significant impact on shaping the region. Today the historic buildings, runways, and anti-aircraft guns are the visible monuments of Midway's aviation past; under the waves are monuments to its naval significance. The USS *Macaw* rests at the bottom of the atoll's channel.

So unless an iconic relic, like the USS *Monitor*, is in danger of being consumed by the corroding effect of salt water, archaeologists working in the national marine sanctuaries leave all parts of the undersea narrative intact.

Only three of the national marine sanctuaries—Monitor, Thunder Bay, and Mallows Bay–Potomac—focus primarily on the nation's maritime heritage, but nearly all the sanctuaries contain historical shipwrecks or archaeological artifacts. Maritime archaeology is an important part of the sanctuary system's work. Congress instructed NOAA to survey, identify, and inventory each sanctuary's historical, cultural, and archaeological sites, as well as significant fossil finds. Each sanctuary cares for its marine artifacts and shares their stories of society's cultural connections to the waters, while the entire sanctuary system's maritime heritage program works nationwide to locate, study, document, protect, and interpret the things found beneath the water.

To tell the story of each artifact, scientists must often begin with the geological context: the changes that have happened to the land and sea over time. The sanctuaries include places that were dry land thousands of years ago: coastal plains, marshes, and forests crossed by streams and rivers.

Off the Northern California coast at what is now Greater Farallones National Marine Sanctuary, the delta of a great river fanned out from a small range of coastal mountains. Ten thousand years ago, water engulfed the delta, turning those mountain peaks into islands. Over thousands of years, seawater slowly rose up the river valley. Eventually, the ocean poured through a gap in the next range of hills to flood another low, open valley. The coastal mountains' peaks are now the Farallon Islands. North of the mountains, what was once an elevated plain lined by bluffs is now Cordell Bank. The gap in the coast range carved by that river is the Golden Gate, and the valley that lay behind it is now San Francisco Bay.

In the Gulf of Mexico and on the Atlantic, coastal lands vanished beneath the sea. Today's Flower Garden Banks National Marine Sanctuary may once have been exposed above the water. In the Florida Keys, archaeologists uncovered remnants of a pine forest buried below the waves off Key West. At Stellwagen Bank, off the coast of Massachusetts, fishers recovered ancient mastodon teeth, proof that land-dwelling animals once inhabited the bank.

These now-submerged landscapes were settings of abundant life. Native Americans once roamed, hunted, and fished them. Beneath the waves at Gray's Reef National Marine Sanctuary, off the Georgia coast, archaeologists found hunters' projectile points. Belief systems around the planet reflect the spiritual power of the ocean as the place where human life began, and where the spirits of the dead travel and come to rest.

In North America, the oldest signs of an oceangoing culture are mounds of discarded abalone shells and stone tools found on Santa Rosa, 26 miles off the California coast and surrounded by Channel Islands National Marine Sanctuary. This is also the site where North America's oldest known human bone

Hōkūle'a ("Star of Gladness") is a modern replica of a traditional Polynesian voyaging canoe designed by Herb Kawainui Kāne, one of the founders of the Polynesian Voyaging Society. On March 8, 1975, *Hōkūle'a* launched from the sacred shores of Hakipu'u-Kualoa; its voyage helped mark a generation of renewal for Hawaii's indigenous people.

Marvels of Seamanship

The ancestors of today's Polynesian peoples sailed thousands of miles across the Pacific Ocean without compasses, sextants, or Western-style charts. Some of their descendants today practice oceangoing navigation techniques, handed down from teacher to student, that rely on feats of memory and observation. Among their tools are a detailed knowledge of weather, winds, and currents and a 16-point mental map of the ocean marked with the positions of more than 200 stars, which change with the seasons and the observer's distance from the Equator. They also deduced the direction of travel by the rhythm of waves striking the boat's hull and could find an island in a seemingly empty sea by reading the patterns of land-refracted waves, reflections on the undersides of clouds, and the flight patterns of birds.

Nainoa Thompson is one of the Hawaiian navigators who is preserving these ancient techniques. He navigated the 2,700-mile voyage between Hawai'i and Tahiti in *Hōkūle'a*, a modern replica of a traditional Polynesian double-hulled sailing canoe. In his essay "On Wayfinding," he writes that traditional navigators ignore birds that stay at sea for months, such as frigate birds and albatross. "Following these birds will not help you find land," he observes. Instead one should follow "the other type of birds . . . those that sleep on islands at night and at dawn go out to sea to fish. . . . When we see these birds in the day we keep track of them and wait for the sun to get low and watch the bird; the flight path of the bird is the bearing of the island."

fragment, a 13,000-year-old piece of a femur, was found. The shells and tools are a little younger, dating back about 12,000 years.

Ice Age hunters left their traces at Thunder Bay National Marine Sanctuary in Lake Huron. Not long ago, archaeologists found the hunters' stone markers and hunting blinds on ridgelines that once were dry ground, then later were drowned by the lake's rising waters.

Humans have used the water as a highway for tens of thousands of years, navigating coastlines with skin boats, log rafts, and planked craft. Some also ventured out into large expanses of open water. Ancient peoples crossed the ocean from Southeast Asia to Australia some 40,000 years ago. Caves in Indonesia and New Guinea contain 20,000-year-old bones of deep-sea fish, fish hooks, and stone weights, evidence of sophisticated offshore craft capable of harvesting food from the seas. But the supreme expression of maritime craft and knowledge, and of humankind's ceaseless drive to explore the unknown, may be the settlement of the islands of the Pacific Ocean.

Archaeologists believe that most of the Pacific's 25,000 islands were unpopulated by humans until the ocean was explored and settled by coastal Pacific and Southeast Asian mariners beginning some 3,000 years ago. Sailing from Papua New Guinea's Bismarck Archipelago, these people, known as the Lapita culture, landed in the Solomon Islands to the southeast, and from there sailed to Vanuatu and on to Fiji. There the Lapita and their descendants remained for more than 1,000 years. Some 2,000 years ago, their descendants, the Polynesians, became the greatest explorers, navigators, and ocean migrants in human history, traveling across millions of square miles to settle the rest of the Pacific's islands.

They followed currents, navigated by the stars, and intimately knew the ocean they traversed. Although almost no archaeological evidence of their great voyaging canoes has survived, their boats were not only seaworthy but also big enough to carry people, livestock, and plants such as taro, all of which helped spread Polynesian culture far and wide. The last great migrations brought the Polynesians to Hawai'i and other islands within the past thousand years. Some of these voyagers may have reached the shores of North America in their canoes.

Archaeological evidence traces many elements of Polynesian culture to the islands of Samoa, which have been inhabited since at least 1300 BCE. The Samoan people kept up brisk inter-island trade and travel using watercraft. Today the *fautasi*, a longboat derived from the blending of Samoan watercraft and Western whaleboat traditions, is used in competitions. The fautasi can be more than 100 feet long and hold up to 50 rowers who paddle together under the guidance of their captain. Keeping a fautasi on course demands skill, practice, and focus, since one rower's mistake can bring the longboat to a halt.

Today American Samoa's seafaring heritage is celebrated in annual fautasi races on American Samoa Flag Day, April 17. Over a dozen fautasi start practicing in December for the competition. Each fautasi crew competes to uphold

Crews practice traditional *fautasi* (longboat) rowing in National Marine Sanctuary of American Samoa. The fautasi races are the highlight of American Samoa Flag Day, which marks the anniversary of American Samoa becoming a US territory on April 17, 1900.

the honor of their islands and districts, as well as traditional ties to the ocean. Fu'ega Sa'ite Moliga, captain of the longest fautasi, *Manu'atele Matasaua*, explains the races' deeper meaning: "The ocean, the land, the forest: everything goes hand in hand. That's what we like to pass on to our kids in the younger generations. To maintain our identity, that's the most important thing to us."

One of North America's oldest examples of an oceangoing watercraft is the *tomol*, the canoe used by the Chumash people for thousands of years to travel the waters of today's Channel Islands National Marine Sanctuary. Tomols were built from redwood trees that drifted down the coast. Made for fishing and transportation, the tomol linked together coastal and island communities. By the early 1830s, the decimation of the Chumash people halted construction of tomols. It was not until 1976 that the first modern tomol was built.

On September 11, 2004, the tomol named *'Elye'wun*, meaning "swordfish," made the journey from the mainland to the village of Swaxil (at the present-day location of Scorpion Valley). In the crew landing *'Elye'wun* in 2004 were five Chumash youths between the ages of 14 and 22, marking a significant passing on of knowledge and experience to a new generation. Following the 2004 crossing, an annual gathering at Limuw brings together Chumash families and

***Tomol* pullers** cross the Santa Barbara Channel. They pass through Channel Islands National Marine Sanctuary to Limuw (Santa Cruz Island), which once was the site of a thriving Chumash village. The Chumash people were the first inhabitants of the Channel Islands.

friends in September, with tomol paddlers making the long cross-channel journey whenever the weather conditions are favorable. Staff from Channel Islands National Marine Sanctuary have been honored to provide assistance with voyage planning, vessel support, and safety coordination.

Many communities on the shores of sanctuaries were founded as trading ports or strategic outposts. Some, like Alpena, Michigan, in Thunder Bay National Marine Sanctuary, or Gualala, California, on the shore of Greater Farallones National Marine Sanctuary, shipped out the raw natural riches critical to the economic growth of the country. Others, such as the ports of Gloucester and Provincetown, near Stellwagen Bank National Marine Sanctuary, were whaling ports, fishing villages, or cities that processed fish harvests for shipment around the world. Cannery Row in Monterey, California, made famous by John Steinbeck, sits on the shore of Monterey Bay National Marine Sanctuary. Whether the harvest was the sardines of Monterey Bay, the menhaden of the Outer Banks, the cod of New England, the whitefish of the Great Lakes, the sponges of the Florida Keys, the tuna of Hawai'i, or the rockfish and salmon of Greater Farallones, Cordell Bank, and Olympic Coast, these places have long, powerful ties of commerce and culture to the waters and resources.

Thick fog and the strong currents of the Santa Barbara Channel were treacherous to maritime traders and other vessels for centuries. Starting in 1932, the Anacapa Island Lighthouse, located in Channel Islands National Park, helped guide sailors through the precarious waters.

Members of Diving With a Purpose assist researchers from Florida Keys National Marine Sanctuary in surveying the wreck of *Slobodna*, lost in 1887. Diving With a Purpose is dedicated to the conservation of submerged heritage, with a special focus on African slave trade shipwrecks and the maritime history and culture of African Americans.

The landscape connecting shore to sea includes docks, piers, wharves, shipyards, boat basins, canals, and seacoast forts, as well as lighthouses and lifesaving stations built to prevent and respond to shipwrecks. Lighthouses help guide mariners through difficult waters. Sanctuaries have some of the most iconic ones: Carysfort Reef Light, emerging from the shallows off the Florida Keys on wrought-iron legs; Cape Hatteras Lighthouse on the Outer Banks, with the beacon atop its 210-foot-tall tower signaling the treacherous sandbars of the "Graveyard of the Atlantic"; Point Bonita Lighthouse near the Golden Gate, a welcome sight to ships' crews covering the dangerous last miles of a long Pacific voyage. In the 19th and early 20th century, lifesaving stations numbered in the hundreds along the coasts and Great Lakes. Now Coast Guard stations have replaced them, but many remain as ruins or preserved historic sites, such as the lifesaving stations at Drakes Bay in Greater Farallones National Marine Sanctuary and Oregon Inlet Station in North Carolina near Monitor National Marine Sanctuary.

Historical records and published accounts indicate that more than 4,000 shipwrecks rest in or near US sanctuary waters. They are crafts of every kind—sailing ships, steamers, fishing boats, tugs, freighters, tankers, and even warships. Some lie close to shore, mangled and mingled with rock and kelp. Others are embedded in coral reefs or buried beneath sand and mud. Still more lie in deeper waters. Some are substantially intact, masts rising toward the surface, cargo still in their holds, cabin doors ajar.

Certain sanctuaries encompass notoriously dangerous seaways. Florida Keys National Marine Sanctuary, with its near-continuous line of shoals and coral reefs, is one of them. From the 1500s through the 1700s, Spanish *flotas*, or fleets, sailed along the Keys carrying looted silver and gold to Spain. Hurricanes sank several Spanish vessels or drove them onto the reefs. But not all the ships lost here carried glittering cargo. Among the wrecks were merchant and military ships bound for the strategic port of Key West astride the Florida Straits, as well as workaday vessels devoted to sponge diving, turtling, and fishing for grouper, snapper, and other fish. These wrecks are tangible reminders of the Bahamians, Afro-Caribbeans, Spanish Cubans, and others who worked in Keys waters.

Near San Francisco's heavily traveled Golden Gate, Greater Farallones National Marine Sanctuary holds nearly 100 wrecks of fishing boats and small two-masted wooden schooners that worked the coastal trades, ferrying lumber, cut stone, hay, produce, and butter and cheese from local sawmills and ranches to sell in the city. Beneath the waves lie the wrecks of ships that had nearly completed their long journeys across the Pacific; either powerful ocean swells swept them onto central California's rocky shore, or they fell afoul of fog, squalls, or simple bad luck while approaching the Golden Gate narrows.

Among the oldest known wrecks in sanctuary waters is the Spanish galleon *San Agustín*, laden with silks, porcelain, and beeswax and lost off the California coast in 1595. Sailing from Manila in the Philippines to Acapulco, Mexico, it stopped to explore the California coast for a potential safe harbor. The crew survived the sinking and reached Mexico in the ship's launch. The wreck now lies buried under surf and sand in the protected waters of Greater Farallones National Marine Sanctuary and Point Reyes National Seashore.

Perhaps the most unusual wreck in the sanctuary system, USS *Macon* was a lighter-than-air, aircraft-carrying dirigible launched by the US Navy in 1933. The 785-foot-long craft was made of alloyed aluminum and fabric, kept aloft by fabric cells filled with helium, and powered by gasoline engines with propellers. *Macon* housed up to four Sparrowhawk reconnaissance biplanes that swung out of their hangar on a metal-framed trapeze. Returning pilots slowly approached *Macon*, hooked onto their trapezes, and were winched back into the airship. *Macon* was lost at sea during a storm off Point Sur, California, on February 12, 1935, when strong winds tore off the upper tailfin, and bits of the wreckage ruptured some of the gas cells. The airship settled onto the sea and sank; two crew members of the 83 aboard lost their lives. In 1991 the Monterey Bay Aquarium Research Institute discovered the wreck, with its brightly painted biplanes still on board, in about 1,500 feet of water inside Monterey Bay National Marine Sanctuary.

Olympic Coast National Marine Sanctuary lies on a heavily traveled sea road: the approach to the Strait of Juan de Fuca and the bustling ports of Puget Sound and British Columbia. Based on a review of literature, more than 180 shipwrecks have been documented in the vicinity of Olympic Coast. Vessels built for both long-distance and coastal voyages have shipwrecked there. By a strange twist of oceanic and historical circumstance, the sanctuary likely also holds the bones of Japanese fishing vessels and coastal traders more than 4,500 miles from their home waters. In 1636 Japan's ruler forbade the construction of any vessels for long-distance travel. But it was not unheard of for Japanese boats and ships to get caught in the powerful Korushio Current—also known as the Black Current or the Current of Death—which flows from Japan to the Pacific Northwest. With no way to escape the current and lacking supplies for a long journey, the crews on these ships often died of starvation, dehydration, or exposure while their vessels sailed on towards Washington State. In 1834 a Japanese coastal trading vessel, *Hojun Maru*, ran aground at Cape Flattery, Washington, with three survivors. Between 1630 and the mid-19th century, there were reports of Japanese ships found in Pacific Northwest waters with only skeletons aboard.

The waters now protected by sanctuaries and monuments have been critical battlegrounds in wars, and their wrecks reflect that history. Hawai'i held strategic importance during World War II. Wartime ship traffic was heavy around the islands, especially near Oahu's famed Pearl Harbor, and fighting was fierce. As many as 80 wrecks of warships, barges, tugs, and landing craft and a large

A kayaker paddles among the skeletons of the "Ghost Fleet" of Mallows Bay–Potomac River National Marine Sanctuary.

number of downed aircraft lie on the seabed in Hawaiian Island Humpback Whale National Marine Sanctuary.

American military vessels also lie wrecked in Papahānaumokuākea Marine National Monument. In 1870 the Navy's Pacific steamship USS *Saginaw* ran aground on a Midway Atoll reef. Remarkably, nearly all of the 93 crew members survived for more than two months until rescuers arrived. In June 1942 hundreds of aircraft were shot down in the World War II Battle of Midway, a decisive US victory that left behind the wreckage of four Japanese aircraft carriers and one American carrier. The atoll is now a national wildlife refuge where millions of seabirds nest.

Many sanctuary shipwrecks are lost until some combination of determined searching and sheer luck reveals their whereabouts. Ships often disappeared without a trace. Historical records can give searchers a starting point for locating shipwrecks, but their information is rarely precise. In the sanctuaries, the work of undersea archaeological discovery often starts with a survey that uses special instruments mounted on ships or towed behind them. Sometimes the survey is carried out for other reasons; for example, sonar mapping of the seabed is done primarily to help ensure safe navigation, but it also reveals the

NOAA divers inspect the USS *Schurz* near Monitor National Marine Sanctuary. The USS *Schurz*, formerly the German cruiser SMS *Geier*, is the only German Imperial Navy warship captured by the US Navy during World War I. The *Schurz* sunk after colliding with the SS *Florida*.

outlines of sunken ships and aircraft. Marine archaeologists also deploy magnetometers. These sensitive devices can detect subtle changes in Earth's magnetic field, like the pull of a huge steel shipwreck lying on the bottom, or the iron spikes, anchors, or cannon scattered around the remains of a wooden ship buried in sand or mud.

Surveys reveal only part of the story. The next step is to dive to get a better look. Archaeologists and volunteers suit up and inspect, map, photograph, and film the underwater find. Investigating the wreckage of ships or planes is like detective work. Archaeologists want to know how old the wreckage is, what it is, what the crew was doing when the ship was lost, and how it was lost. The goal is to gather as much information on the wreck as possible and prepare a detailed record for ongoing study.

Shipwreck diving is magical—and risky at almost any depth. Shipwrecks gather layers of silt that, once disturbed, can cloud visibility in seconds. Divers can become disoriented, entangled, entrapped, or hurt by sharp objects that can damage scuba equipment. And at depths greater than about 100 feet, breathing air from a tank can bring on nitrogen narcosis. Also known as the "martini effect" because its effects on the brain are akin to drinking several martinis, nitrogen narcosis can severely impair divers' judgment. Below about 150 feet, divers must use specially mixed gases that contain helium to displace nitrogen in the breathing mixture.

In even deeper water, robots do the work. Untethered and free-swimming robots are called autonomous underwater vehicles, or AUVs. Launched from survey ships, they can go deep to conduct high-resolution sonar surveys, which are used to make three-dimensional maps. This is the technology that sanctuary system archaeologists and their partners used to map the wreck of *Titanic* under two and a half miles of water in the North Atlantic in 2010. Other robots, ROVs, are tethered to a surface survey ship, where they work in depths ranging from a few hundred feet to miles below the surface.

ROVs were used to explore the wreck of the passenger steamer SS *Portland* near Stellwagen Bank, off Boston, Massachusetts. *Portland*'s sinking was one of the most tragic shipwrecks in New England's history. The ship was an elegant coastal steamer carrying passengers between Portland, Maine, and Boston. On November 26, 1898, the steamship sailed from Boston despite storm warnings. Caught in a torrential gale that sank more than 150 vessels, *Portland* was lost with no survivors. The gale took more than 400 lives, including *Portland*'s estimated 192 passengers and crew. The exact number will never be known since the passenger list went down with the ship.

The steamer's final resting place was a mystery until shipwreck explorers John Fish and Arne Carr found it in 1989, lying in over 400 feet of water in Stellwagen Bank National Marine Sanctuary. When a NOAA expedition returned to the wreck in 2002, a camera on an ROV revealed that the steamship

An ROV investigates the boiler and condenser of the wooden steam barge *Montana*, which sank in 1914 in what is now Thunder Bay National Marine Sanctuary.

NOAA maritime archaeologist Kelly Gleason displays one of five whaling harpoon tips discovered at the *Two Brothers* shipwreck site.

The Wreck of *Two Brothers*

If ever a ship's captain deserved to be called ill-fated, it might be George Pollard, the last master of the whaling ship *Two Brothers*. Early 19th-century whalers were essentially the factory ships of their era, and hundreds of them journeyed to Hawai'i in the 1820s. *Two Brothers* was not Pollard's first command. In 1820 he was captain of the whaler *Essex* when a whale rammed into and sunk the vessel. Pollard and other survivors lived through a hellish ordeal, drifting in open boats under a blazing sky for 95 days and eating the flesh of dead comrades. Their story inspired Melville's *Moby-Dick* and was the subject of the bestseller *In the Heart of the Sea* by Nathaniel Philbrick.

It was *Two Brothers* that rescued Pollard and his crew. When *Two Brothers* returned to its home port of Nantucket in 1821 with the *Essex* survivors, Pollard was given command of the ship that had saved him, and so returned to whaling.

Pollard was shipwrecked again when *Two Brothers* ran aground on French Frigate Shoal in the Northwestern Hawaiian Islands on February 11, 1823. He was also rescued again, along with his crew. But the wreck of the whaling ship remained undiscovered, slowly merging with the coral and sand until 2008, when a NOAA team surveying Papahānaumokuākea Marine National Monument found it. The wreck of *Two Brothers* is the only tangible link to the events that inspired Melville's classic novel. *Essex* lies deep in an unmarked grave in the middle of the Pacific.

was festooned with fishing nets that, over the years, had snagged on the unseen wreck. The doomed passenger ship lay upright and almost intact, a coffee mug still in place on the deck. The "*Titanic* of New England" remains entombed in the deep ocean darkness, awaiting further exploration.

ROV surveys have examined wrecks in much deeper water. In 2016 the Ocean Exploration Trust carried out the first survey of the World War II aircraft carrier USS *Independence*, which lies in 2,800 feet of water off the California coast in Monterey Bay National Marine Sanctuary. Sanctuary system archaeologists also use human-piloted submersibles to explore shipwrecks.

These new technologies are efficient and return significant scientific data. ROV-carrying research ships, such as NOAA's *Okeanos Explorer*, use satellite links to livestream their discoveries, thus opening up the hidden world of undersea exploration to the public and making it possible for interested scientists to remotely participate in the work. Dozens or more scientists from different disciplines, ranging from oceanography, marine biology, and geology to archaeology and history, cooperatively direct the collection of data on currents, oxygen levels, and temperatures and the sampling of deep-sea water, and analyze the results. Researchers have discovered many new life forms in these water bottles from the deep.

In very rare cases, archaeologists excavate wrecks in sanctuaries or collect artifacts from them. The best-known example is the Civil War warship USS *Monitor*, the Union ironclad that has endured trawl impacts, anchor snags, and severe corrosion. Because the ship was such an icon, Congress instructed NOAA and the Navy to recover key parts of the wreck and bring them ashore for preservation and display. Teams of experts worked for several years to raise the engine, propeller, anchor, massive turret with its two cannons, and many other artifacts. Also recovered were the skeletons of two of the 16 men who were lost when the ironclad sank. Some of the ship's components are being treated to stop the corrosive effects of more than a century at the bottom of the sea. Others are displayed at the USS *Monitor* Center at the Mariners' Museum in Newport News, Virginia.

The sanctuaries' seafloors, from the tip of Cape Cod to American Samoa's Fagatele Bay, tell the story of how the ocean and the Great Lakes surround, nurture, and sustain humankind. They are places to discover our country's maritime heritage—a legacy of thousands of years of exploring, harvesting the seas' bounty, making perilous journeys to new lands, building coastal communities, and making good use of maritime traditions, sea lore, and sea wisdom. By studying, protecting, and promoting this diverse legacy, sanctuaries and monuments help us understand our past. ○

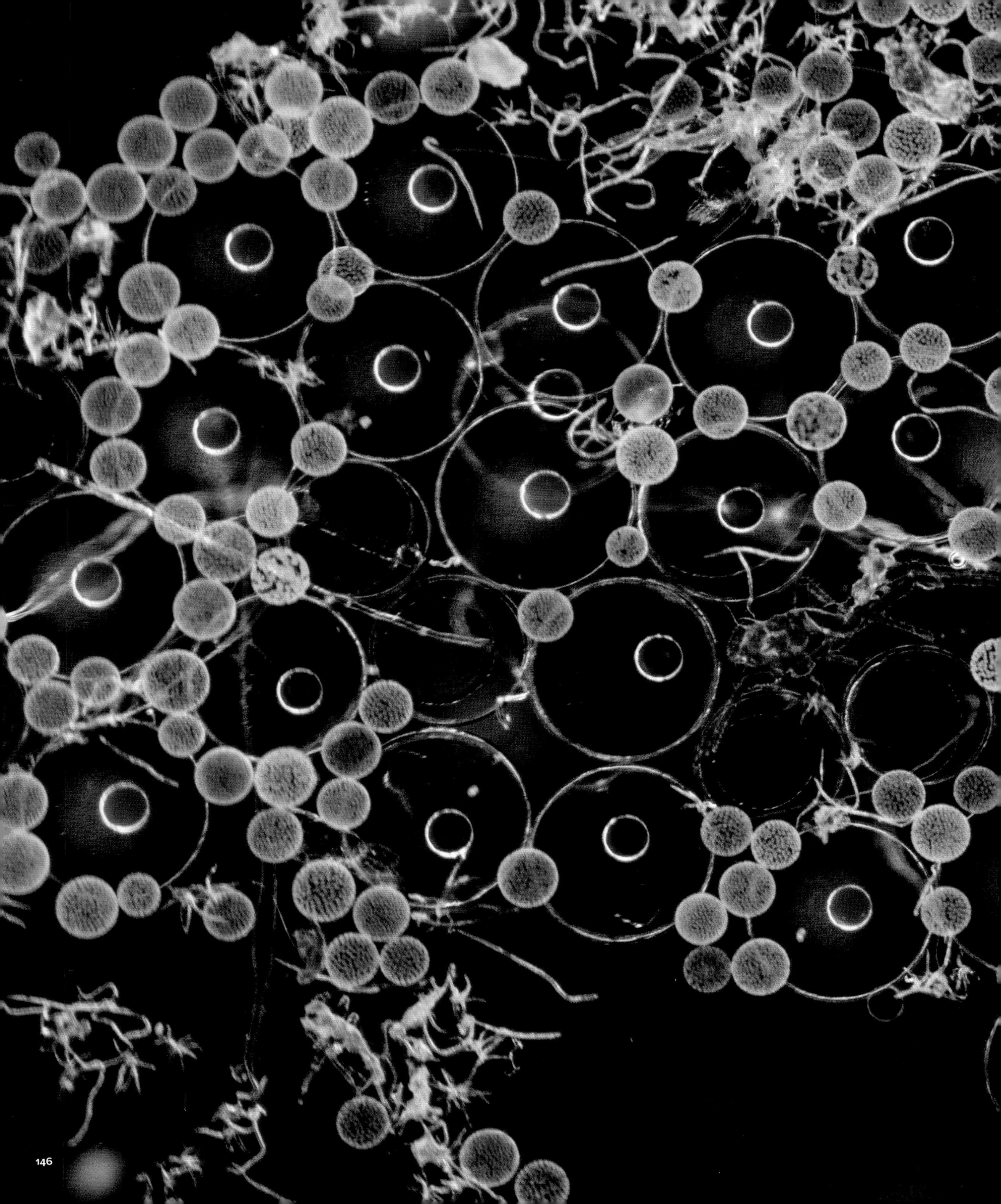

4

OCEAN WEALTH

In every outthrust headland, in every curving beach, in every grain of sand there is the story of the earth.

RACHEL CARSON

THE OCEAN, GREAT LAKES, AND THEIR BOUNTY are essential to US and global economic health and security. This is as true today as it was in our past. For most of the United States' history, the ocean was viewed strictly in terms of its economic usefulness: as a source of food, a means to transport goods to and from markets, and a bulwark against America's enemies. Today we increasingly recognize how vital it is to our ecological security and well-being.

The ocean economy is big, fast-growing, and complex, and like the ocean itself, it spans international borders. The US government includes six groups of industries in the ocean economy: living marine resources such as fish, shellfish, and seaweed; shipping; boat- and shipbuilding; marine construction such as dredging and beach nourishment; offshore minerals extraction; and tourism and recreation. In 2016 the US ocean economy contributed an estimated $304 billion to the national gross domestic product and provided 3.3 million jobs, nearly three-quarters of them in tourism and recreation. The international Organization for Economic Cooperation and Development (OECD), whose 36 member nations include the United States, estimated "very conservatively" in 2016 that the ocean adds $1.5 trillion per year to global wealth.

Another way to look at the economic worth of the ocean is to focus not on what humans can take from it and sell, but what it provides us free of charge. There are things nature does for humankind that are essential to our survival, or that vastly improve our quality of life. Resource economists refer to these as ecosystem services. Some ecosystem services would be extremely costly to replace. Others can't be duplicated by humankind at any price. The benefits the ocean provides us are so fundamental that it's difficult to imagine life on Earth without them.

Between half and four-fifths of the oxygen in Earth's atmosphere comes from microscopic algae in the ocean, which photosynthesize just like terrestrial plants—they take in carbon dioxide and emit oxygen, ensuring that we humans have a breathable atmosphere. The ocean is crucial to the continuous

Mangroves line more than 1,800 miles of shoreline within Florida Keys National Marine Sanctuary. Mangrove forests stabilize the coastline, reducing erosion from storm surges, currents, waves, and tides. The intricate root system of mangroves also makes these forests attractive to fish and other organisms seeking food and shelter from predators.

PREVIOUS PAGES
Larval fish eggs and green algae float in the waters of Stellwagen Bank National Marine Sanctuary. Early European fishing crews were drawn to the bank because of the abundance of fish, which are a vital component of the sanctuary's biological diversity;

natural cycle of water on Earth: all streams and rivers flow to the sea, and as water evaporates from the sea, it gives rise to life-sustaining rain and fresh water. Ocean fish, shellfish, and even seaweeds are essential sources of nutrition for more than 7 billion people. And as our climate changes, with human activities like the burning of fossil fuels causing levels of carbon dioxide in the atmosphere to soar, the oceans are naturally acting as what scientists call a carbon "sink," absorbing and storing some of the carbon that would otherwise enter the atmosphere.

Nature provides natural infrastructure that protects coastal communities and improves water quality. When storms strike the coast, underwater features such as coral reefs, coastal wetlands, banks, and gently sloping continental shelves slow down or break up storm waves and reduce the extent of flooding and damage on land. Healthy coral reefs can reduce wave energy by 97 percent. Mangroves, salt marshes, underwater sponge gardens, shellfish, and wetlands filter nutrients and sediments out of water, keeping it cleaner for people and wildlife. A single oyster can filter 180 liters of water every day. Even the ocean's beauty benefits humankind, inspiring art and literature, offering us opportunities for exercise and play, giving us a sense of tranquility and timelessness.

How do you put a monetary value on these intangible but vital benefits? Economists have ways to calculate the dollar values of natural resources and the benefits they provide. For example, to estimate the worth of commercial fishing, they calculate the sales of the fish themselves; the income that people earn in the fishing industry; and the value of all the goods and services that keep commercial fishing boats afloat, get their harvests to market, and keep the tables turning in seafood restaurants. Other assessments involve the "use value" of a resource—how much a person pays to go charter fishing or scuba diving—and the "non-use value," how much a person would be willing to pay to protect a coral reef or a whale. To calculate the monetary value of ocean habitats or living creatures, like coral reefs or schools of fish, economists assume these natural resources have been damaged or lost by accidents like oil spills or ship groundings. They calculate what it would cost to restore the damage or rebuild the population, as well as the value of the resources' "lost use" during the time it takes to restore them.

There are no perfect methods for calculating the value of the ecosystem services. It's clear that estimates of ocean ecosystem services' worth do not capture the full value of what the oceans, including sanctuaries and monuments, provide to the planet. Looking only at the services provided by coral reefs in the United States, a 2013 study estimated the reefs' total economic value was a whopping $3.4 billion per year. One study set the value of commercial and recreational fishing in the eastern tropical Pacific, which stretches from Southern California to Peru, at nearly $4 billion. In that same swath of the ocean, just one ecosystem service—carbon storage—has been valued at more than $12 billion.

Kayakers share Monterey Bay National Marine Sanctuary with squid fishing boats, typically California's most valuable and productive fishery. In Monterey, fishing and outdoor recreation are important to the local economy.

While we may not be able to place a definitive dollar sign on the worth of marine sanctuaries and monuments, we can understand some of their benefits. Healthy ecosystems in sanctuaries, such as coastal forests, seagrass beds, and salt marshes, trap and store carbon dioxide, removing it from the atmosphere. The roots of mangroves and coastal forests protect shorelines from erosion. Underwater shellfish beds and coral reefs block waves and storm surges and slow them down. Salt marshes and other coastal wetlands help absorb floodwaters, and sand dunes protect against flooding.

Sanctuaries protect the economic benefits that nature provides. For example, sanctuary managers sometimes work with federal and state fishery management agencies to limit where, when, and how seafood is harvested. The result is often an increase in the number, size, and variety of fish and shellfish, both inside and outside no-fishing zones. By protecting breeding populations of fish and shellfish, the sanctuaries' catch limits actually increase the seafood catch.

To effectively conserve each sanctuary's assets, it is vital to have information about the health of its wildlife, habitats, natural processes, and archaeological or cultural resources such as shipwrecks. NOAA experts prepare condition reports that now include descriptions of the ecosystem services each sanctuary

Surfers look seaward at dawn from a beach at Olympic Coast National Marine Sanctuary. Locals and a growing number of surfers from Seattle and beyond have discovered the challenges and rewards of Olympic Coast breaks, which are fueled by big Pacific swells.

provides—economic benefits such as clean air, fresh water, and food; harvested sea life such as fish, shellfish, plants, and ornamental items; medicines or chemicals derived from marine plants or animals; energy; clean water; and coastal protection. The condition reports, available online, give managers a foundation for decisions about how to sustainably manage each sanctuary.

For scientists, sanctuaries offer research opportunities. Especially important is the chance to work at research sites that they know they can return to for years. In Florida Keys National Marine Sanctuary, some of the area's original coral reef scientists set up research sites and established studies that have now existed for 50 years or more, and the elder scientists have handed down this work to a new generation of researchers. These long-term studies offer unparalleled opportunities to understand how changing conditions are affecting the reefs' health. By better understanding changes on the reefs, sanctuary managers gain the information they need to better protect these resources, now and in the future.

The sanctuaries' wildlife populations also support ocean tourism and recreation, with all that it contributes to the ocean economy and to our quality of life. Surveys show that one out of every seven Americans spends recreation time on the water. Kayaking, windsurfing, scuba diving, surfing, sailing, and stand-paddle boarding are all popular, and each year more than 10 million Americans go fishing. More than 40 accredited aquariums host millions of visitors each year, a testament to Americans' fascination with the wonders hidden under the sea. Half of the nation's fastest-growing outdoor activities are water based. Today, Americans are spending even more time on, in, or near, water.

Some of America's most popular water sports and ocean getaway spots are in national marine sanctuaries. Whale watching is famously good at Stellwagen Bank and Hawaiian Islands Humpback Whale national marine sanctuaries. Surfers from all over the world seek out well-known surf spots in the waters of Hawaiian Islands Humpback Whale National Marine Sanctuary. The Banzai Pipeline and Waimea Bay on Oahu's famous North Shore are just two of the sanctuary's surf meccas. Surfers also throng to California's Monterey Bay National Marine Sanctuary, testing their skills against the waves at Mavericks off Half Moon Bay, Steamer Lane off Santa Cruz, and many other spots. Several sanctuaries are famed destinations for scuba divers. Most lists of the top places to dive in the United States include Monterey Bay, Channel Islands, Florida Keys, Flower Garden Banks, and Thunder Bay sanctuaries.

Nature-loving families visit the shores of Monterey Bay, Greater Farallones, and Olympic Coast sanctuaries, usually when the tide is low, to explore the bountiful and beautiful sea creatures that inhabit virtually every small indentation in rocks and sandbars. The same sea stacks that provide toe-level tide pool habitat for colorful starfish, anemones, and other creatures are also teeming overhead with seabirds nesting, feeding, squabbling, and fledging, making

National marine sanctuaries provide tremendous recreational opportunities for visitors, from fishing to boating to swimming. A visitor to the Florida Keys National Marine Sanctuary fly fishes in its turquoise waters. Recreational fishing in the sanctuary is worth more than $270 million annually.

these sanctuaries great destinations for birders wearing binoculars, as well as for tide poolers peering through hand lenses.

At Florida Keys National Marine Sanctuary, history buffs and undersea adventurers can explore the Florida Keys Shipwreck Trail, a series of nine sunken vessels ranging from a Spanish treasure ship that sank in a hurricane in 1733 to a World War II troop ship that, once its service was ended, was deliberately sunk to form an artificial reef. The Great Lakes Maritime Heritage Trail has breathtaking vistas and information about the maritime history of Thunder Bay National Marine Sanctuary. And the Mallows Bay–Potomac National Marine Sanctuary offers opportunities for kayakers and canoers to paddle among the starkly beautiful hulls of its "Ghost Fleet."

For people who live and work near the national marine sanctuaries, the benefits of these protected places are apparent. There is a special term for the cities and towns that are the jumping-off places for visitors to a national park or a sanctuary: gateway communities. Typically, they offer places to eat and spend the night; opportunities for tours, boat trips, and other kinds of recreation; and often aquariums, museums, or interpretive centers that add new dimensions of learning or play to the experience.

The gateway communities for the national marine sanctuaries are as varied as the sanctuaries themselves. Some, like Pago Pago, are small communities perched between lushly forested landscapes and undersea Edens. The capital of American Samoa and the gateway to the only national marine sanctuary south of the Equator, Pago Pago is built around a scenic natural harbor and has a population of about 4,000.

In Hawai'i, whale watching brings tourists from across the globe to catch a glimpse of ocean giants. Many trips take place in the waters of Hawaiian Islands Humpback Whale National Marine Sanctuary. From the sanctuary's beachfront visitor center in Kihei on the south shore of Maui, visitors can sometimes see humpback whales breaching, and can learn about these iconic marine mammals, along with other native Hawaiian ocean creatures such as sea turtles and monk seals.

On the mainland, the sanctuaries' gateway communities are strung like small gems along the coasts, inviting visitors to explore, play, and learn. A road trip from one gateway community to the next might begin across the vast Pacific on the Washington State coast, where the town of Port Angeles faces the Strait of Juan de Fuca, with its back to the Olympic Mountains. Near the ferry pier sits the Olympic Coast Discovery Center, where visitors can learn about marine conservation, the local sea life, and scientific exploration in the sanctuary waters. In the sanctuary, visitors can fish, sea kayak, surf, and dive. Charters for salmon, halibut, lingcod, and tuna fishing are available from Neah Bay, La Push, Forks, Sekiu, and Westport. And hikers to Cape Flattery, on the Makah Indian reservation, can walk to the northwesternmost point in the lower 48 states.

A long drive down the coast passes through Point Arena and the artist enclave of Gualala in California's Sonoma County, along the shore of Greater Farallones National Marine Sanctuary. Visitors can gallery-hop through the town, hike the dunes of an extensive oceanfront park, and golf at the top-rated Sea Ranch Golf Links. The nearby community of Olema, California, in Marin County, perches on the edge of Point Reyes National Seashore and serves as the headquarters for Cordell Bank National Marine Sanctuary. The town's name is drawn from the Miwok word for "little coyote." More than 100 Miwok sites have been identified on the Point Reyes Peninsula, and the town features a reconstructed Miwok village.

From there, it's a short trip south and across the Golden Gate Bridge to San Francisco, named for the patron saint of animals and ecology. This great American city is the headquarters for Greater Farallones National Marine Sanctuary, which lies just beyond the fabled Golden Gate. The sanctuary's visitor center is just yards from the San Francisco bayfront in a historic Coast Guard station at the Presidio, a former Army base that is now a national park, with views of the Golden Gate Bridge.

About 120 miles south on the spectacular Pacific Coast Highway lies Monterey, home of the Monterey Bay Aquarium, Cannery Row, and Monterey Bay National Marine Sanctuary. The sanctuary's Exploration Center is steps away from the ocean and the wharf in Santa Cruz, a lively university town and a surfer's paradise. Built in 2012, the center is a model for sustainable green design and has multimedia exhibits to help visitors explore the sanctuary. Further south on San Simeon Bay is the Coastal Discovery Center, a joint venture between Monterey Bay National Marine Sanctuary and California State Parks. The center celebrates the connection between land and sea.

The region around Monterey Bay National Marine Sanctuary is a unique hub for ocean science, based on the critical marine habitats and wildlife, long

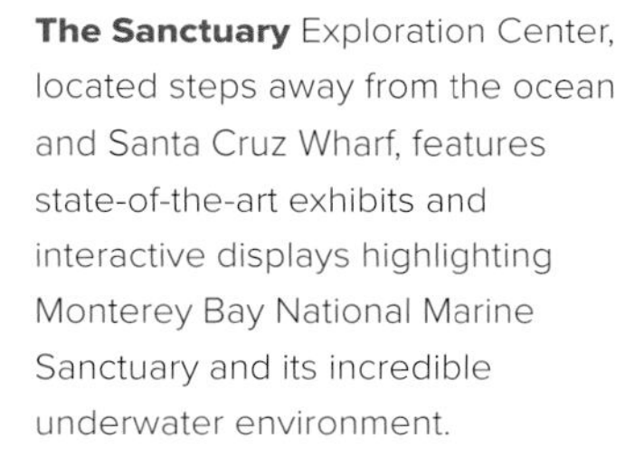

The Sanctuary Exploration Center, located steps away from the ocean and Santa Cruz Wharf, features state-of-the-art exhibits and interactive displays highlighting Monterey Bay National Marine Sanctuary and its incredible underwater environment.

history of research, and the presence of a number of academic institutions. In 2016 the area was home to 24 marine science facilities employing more than 2,000 scientists and support staff, with annual budgets totaling more than $337 million.

The last Pacific Coast stop is Santa Barbara, located along one of the few south-facing sections of the California coast and backed by the dramatic Santa Ynez Mountains. When the first Spanish explorers arrived here in the early 16th century, 8,000 or more Chumash people lived along this coast; today the area's population totals about 92,000. The Channel Islands National Marine Sanctuary office can be found on the bluff tops of the University of California, Santa Barbara. A visitor's center, boating center, and maritime museum all offer Santa Barbara visitors ways to explore the sanctuary.

The Gulf Coast barrier islands are hot spots for tourism. Galveston, Texas, has been a hub of tourism and shipping since the 1830s. Once known as the

Major shipping lanes converge in Greater Farallones National Marine Sanctuary. Ship strikes of whales are a continuing concern for all West Coast sanctuaries.

YANG MING
K LINE
K LINE
K LINE
MOL MAGNIFICENCE

"Playground of the South," it was devastated by a hurricane in 1900 but gradually rebounded and is now a major cruise port. The town is rich in history and is a mecca for birding. In Galveston and nearby towns along the South Texas coast, commercial dive and fishing charter boats take visitors out to Flower Garden Banks National Marine Sanctuary.

Mile Marker 0 begins the Atlantic Coast sanctuary tour, as well as the start of the Overseas Highway in Key West, Florida, and the southernmost point in the continental United States. "Key West" is a corruption of the Spanish phrase *Cayo Hueso* ("Bone Island"), and the stories of the name's origin are as many and varied as the town's lively festivals and parades. The turquoise waters of Florida Keys National Marine Sanctuary run the length of the 220-mile tourist-friendly island chain. Florida Keys National Marine Sanctuary has a huge tourism industry that centers on diving, snorkeling, boating, fishing, and enjoying the culture and history of the Keys. The National Marine Sanctuary Foundation conducted a study in 2019 that found the Florida Keys National Marine Sanctuary contributes $4.4 billion annually to the state's economy.

On the Florida mainland, the Overseas Highway becomes US 1, running up the Atlantic Coast from Key West to Maine. The next sanctuary stop is Savannah, the oldest city in Georgia, famed for its beautiful public squares and gardens, and entry to Gray's Reef National Marine Sanctuary. From nearby Tybee Island, captains offer charters for rod-and-reel fishers or experienced ocean divers, while the Tybee Marine Science Center is one of seven partner sites, stretching from Columbia, South Carolina, to Atlanta, Georgia, where visitors can learn about the sanctuary.

Heading north up the coast from Savannah, travelers arrive at historic Newport News, Virginia, the headquarters of Monitor National Marine Sanctuary and the home of the USS *Monitor* Center. Here, visitors can see how the most important artifacts retrieved from the sunken Civil War ironclad ship *Monitor* are being conserved, view a replica of the famed warship's gun turret, and learn about the historic battle of two ironsides, *Monitor* and CSS *Virginia* (known before the Civil War as USS *Merrimack*, before it was retrofitted and rechristened by the Confederacy).

History buffs can follow Interstate 95 north toward Nanjemoy, Maryland, to Mallows Bay–Potomac River National Marine Sanctuary. From the ironsides of the Civil War to the wooden steamships that make up the "Ghost Fleet" of World War I, national marine sanctuaries offer visitors to these waters opportunities to learn about how the ocean and Great Lakes shaped cultures and economies. Today Mallows Bay–Potomac River is a popular spot for canoeing, kayaking, fishing, birdwatching, and other outdoor recreation.

Mallows Bay is an outdoor classroom and living laboratory for research, conservation, and learning opportunities because of its distinctive maritime features and connection to the Chesapeake Bay. The National Marine Sanctuary

Blue Star dive operators in the Florida Keys are participating in efforts to remove underwater marine debris that can injure corals, damage sponges, and entangle wildlife.

Sea Stars

The Blue Star program, established at Florida Keys National Marine Sanctuary, recognizes dive and charter boat operators committed to sustainable diving, snorkeling, and fishing practices. To earn Blue Star certification, companies must train staff in practices that reduce visitors' impact on sensitive habitats. They must promise to use the sanctuary's network of mooring buoys instead of risking coral injury by dropping their anchors on coral habitat, and they must commit to supporting a conservation-related project. The program gives owners valuable training and a marketing opportunity, and it benefits the sanctuary by reducing visitor impact. Sanctuary staff helped other facilities develop or manage similar recognition programs, including Dolphin SMART for wildlife viewing tours in Florida, Alabama, and Hawai'i; Whale SENSE for whale-watching cruise operators at Stellwagen Bank and in the Gulf of Maine; and ANCHOR, based at Monitor National Marine Sanctuary, for shipwreck divers.

"There are people who specifically come to us because we are a Blue Star operator," comments the owner of a dive shop in Key West. "We became Blue Star certified because we wanted to be known as a dive store that cared about the environment," says the proprietor of a dive shop in Islamorada. "We love what Blue Star represents," agrees the operator of a water tour and sports business in Key West. "I think everyone that goes to the reef should be Blue Star certified."

A boy explores the rocks along a shoreline. Protecting our shorelines with national seashores, lakeshores, and ocean parks helps ensure that this child's grandchildren will be able to play at the same beautiful spot.

Foundation, in partnership with the Maryland Department of Natural Resources, is supporting a five-year water quality monitoring program at Mallows Bay to provide continual, real-time data to the public, resource managers, and scientists via the Eyes on the Bay program. The partnership will aid commercial watermen and recreational anglers and benefit local tourism operators and visitors by providing real-time data for planning trips.

Heading north, past the urban centers of Washington, DC, New York, and Boston, the East Coast sanctuary road trip ends in Scituate, Massachusetts, a fishing town incorporated in 1636 by English colonists. It sits halfway between Boston and Plymouth and is the headquarters of Stellwagen Bank National Marine Sanctuary. A small museum in Scituate has an exhibit on the tragic sinking of the steamship *Portland* on Stellwagen Bank, while the New England Aquarium in Boston has a bow-front aquarium showcasing a deep boulder reef characteristic of Stellwagen Bank.

Boats from Boston, Plymouth, Gloucester, and Cape Cod still fish commercially on Stellwagen Bank National Marine Sanctuary, as New England fishers have done for more than 300 years. Now those ports also take part in a lucrative whale-watching industry centered on the sanctuary.

The ocean offers powerful lures for visitors: wildlife from sea stars to sea otters, from white abalone to black oystercatchers; habitats ranging from brine seeps to coral reefs; seascapes of transcendent beauty; cultures eons old; and centuries of maritime tradition.

Farthest north, on the shores of Lake Huron, lies the town of Alpena, Michigan. This small, resilient city of 10,000 swells with visitors eager to enjoy Thunder Bay National Marine Sanctuary, its lighthouses, and its rugged, forested islands. There are abundant opportunities for kayaking, canoeing, and diving, and glass-bottomed boat tours give visitors the chance to see the sanctuary's amazingly preserved shipwrecks, without having to don scuba gear or pick up a paddle.

Sanctuaries increase our awareness of the importance of the ocean and Great Lakes to our everyday lives. They offer beauty and encourage healthy outdoor pastimes. Their visitor facilities and outreach programs welcome local residents with a neighborhood approach that can help motivate citizens and communities to practice conservation. The sanctuaries contain not only natural wonders but also cultural and historical treasures: ships and shipwrecks, seafaring traditions and cultures, the relics of an ancient sea, and the life of the present one. By protecting the wildlife, wild waters, and the character of these coastal towns, the sanctuaries help improve the quality of life of their nearby communities. ○

5
STEWARDSHIP

whatever we lose (like a you or a me)

it's always ourselves we find in the sea

E. E. CUMMINGS

MANY AMERICANS' FONDEST MEMORIES are of days at the beach: sprinting through the blistering hot sand toward the cool, refreshing ocean; staring off into the endless blue horizon—the Great Lakes' infinite expanse; hearing the loud swoosh of waves crashing and the rush of foamy white water creeping up the shore; searching for a spot on the blanket-dotted beach; flying a kite along the wide, sandy strand; fishing alongside foraging seabirds and family; marveling at sea stars and sea anemones among the rocky tide pools; or swimming among angelfish in warm tropical waters. Experiences like these inspire people to protect our ocean and Great Lakes.

At national marine sanctuaries, people of all ages and backgrounds can volunteer as citizen scientists, tracking trends in the health of the sanctuaries' resources, or teach others how to enjoy those resources respectfully. They can become researchers, adding to our store of knowledge about sea life. They can speak on behalf of nature in the rooms where management decisions are made, helping to build strong ties between the ocean community and the community on land.

Public stewardship is at the heart of national marine sanctuaries and monuments. Hundreds of volunteers, nonprofit organizations, businesses, and local and state governments work alongside the staff of NOAA's Office of National Marine Sanctuaries, helping to conserve these treasured places for future generations. National marine sanctuaries offer the opportunity to give back to the ocean and Great Lakes that give so much to our society.

For example, scientists share their data, and legal and policy experts provide advice on regulations and programs. Local businesses and residents also play a crucial role in providing real-world perspectives as the sanctuary managers make decisions about how to care for these special places: how will proposed regulations affect the grandparents who take their grandchildren fishing, or the tour guide who wants to show healthy coral reefs to customers?

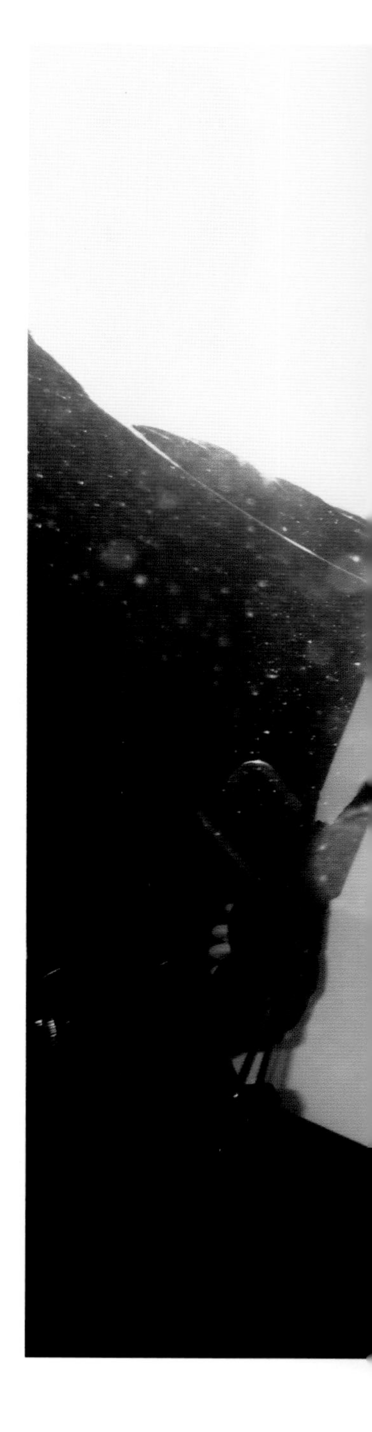

A diver, inspired by Carysfort Reef in Upper Key Largo, at Florida Keys National Marine Sanctuary, celebrates the gift the ocean has given her: freedom.

PREVIOUS PAGES

The Olympic Peninsula's intertidal marine life is diverse. Tide pools are rich with sea stars, anemones, and crabs. Tide pooling is a perfect family activity, allowing people young and old to explore our coasts. Here a family examines the shell of a shore crab.

Community members help answer these important questions by serving on sanctuary advisory councils. People with different perspectives and from different backgrounds advise and support sanctuary superintendents through these councils. The first citizen advisory council was established in 1990 in the Florida Keys. Today each of the sites in the sanctuary system has an advisory council, with a total of more than 400 members serving on these councils. Each council reflects the particular situation of its sanctuary. For example, Cordell Bank National Marine Sanctuary is an offshore site that doesn't get many visitors and focuses on science, education, and resource protection, so its advisory council includes two research members, two education members, two conservation members, two maritime business members, and two fishing members.

Will Benson is a Blue Star fishing guide in the Florida Keys. "Every time my young son, Luke, says, 'Dad, I want to go fishing. I want to catch tarpon,' I'm reminded that I'd better do my part now to make sure that that future is available to him."

The islands alongside Florida Keys National Marine Sanctuary are bustling with visitors, so its advisory council includes members whose fishing, diving, and tourism businesses rely on a healthy ocean.

One member is Will Benson, a fishing guide and filmmaker. Benson grew up fishing and diving in the Keys, and became a professional recreational fishing guide at age 19. The waters on both sides of the island chain are a mecca for sport fishers, who come from all over the world to fish for silvery tarpon, elusive bonefish, and other sought-after species. Through his guide work, films, and advocacy, Benson works to spread sustainable fishing practices. He is building and strengthening partnerships between the recreational fishing community and Florida Keys National Marine Sanctuary, increasing public awareness about the benefits of the sanctuary, and encouraging local businesses and residents to take part in conservation projects. As a member of the Blue Star Fishing Guides, Will is shaping a conservation and education initiative that has the potential to reach thousands of visitors to national marine sanctuaries across the country.

A native Hawaiian born and raised on the island of Kaua'i, Pelika Andrade works with the University of Hawai'i Sea Grant program and serves on the Papahānaumokuākea Advisory Council. She has a master's degree in Hawaiian studies with a focus on *mālama ʻāina* (Hawaiian conservation) and has visited the remote Northwestern Hawaiian Islands many times to conduct scientific and cultural research. Andrade describes her trips to the monument: "My first trip in 2008 to Pihemanu Kuaihelani (in Hawaiian) . . . Midway Atoll (in English) opened my eyes to the extent and expanse of our Hawaiian universe, but more importantly has awoken an awareness and

Pelika Andrade (right) serves on the Papahānaumokuākea Advisory Council. Pelika made a lifetime commitment to *mālama ʻāina* (kai), which she shares with her family and community: "I feel both honored and privileged to be a contributing part of caring for our islands, our elder siblings, and ensuring a healthy-thriving *paeʻāina* (archipelago) for future generations."

responsibility that I have accepted and will pass on to my children and theirs. Our *kupuna* (Grandfather) islands, our realm of *pō* (darkness, the gods), contains countless treasures for us as *kanaka* (human beings) and as citizens of our global community. It is in this realm that we are born from and return to. The Papahānaumokuākea is where our eldest of kupuna reside and continue to thrive, our corals and reef systems, fish, plants, and birds. I feel both honored and privileged to be a contributing part of caring for our islands, our elder siblings, and ensuring a healthy-thriving *paeʻāina* (archipelago) for future generations."

Francesca Koe, a member of Greater Farallones National Marine Sanctuary in Northern California, is a dedicated advocate for the sanctuary. A dive instructor, competitive free diver, and champion for ocean conservation, Francesca was instrumental in designing and implementing California's system of marine protected areas along its coast. She is now tackling the questions: what

Kauaʻi Ocean Discovery in Hawaiʻi is a new educational facility opened by NOAA and the National Marine Sanctuary Foundation to share the traditions and knowledge of our ocean connections and to inspire stewardship in its visitors. Here Hawaiian dancers welcome visitors at the opening of the facility

KAUA'I OCEAN DISCOVERY

is causing declines in California's bull kelp forest, and what should we do to help? She is working with the staff of Greater Farallones National Marine Sanctuary and the California Department of Fish and Wildlife to explore how to build resilience and restore this critical habitat.

The Coastal Treaty Tribes of the Olympic Peninsula—the Hoh, Makah, and Quileute tribes, and the Quinault Indian Nation—view the continued ability to harvest and use water, plants, mammals, fish, and other resources of the region as critical to the protection of their treaty rights and the continuity of their cultures. Through treaties, the US government recognized their tribal sovereignty and certain tribal rights. These include the right to take fish, shellfish, and, for the Makah, whales and marine mammals. The Intergovernmental Policy

Francesca Koe is a dedicated advocate for national marine sanctuaries. "There is so much to discover and explore, beauty to behold, and power to be humbled by. I feel eternally grateful for every chance I get to delve under the surface and to witness the wonder and mysteries of my preferred habitat."

Council, a forum of tribal, state, and federal managers, was created in recognition of this unique relationship, and of the federal government's special "trust responsibility." It enables the participants to discuss issues of mutual concern and increase collaboration to achieve common objectives related to resource management within the sanctuary. In addition to the Intergovernmental Policy Council, the Makah Tribe, Quileute Tribe, Hoh Tribe, and the Quinault Indian Nation all have seats on the Olympic Coast National Marine Sanctuary Advisory Council.

Local voices must be heard when the long-term health of a sanctuary requires managers to take steps that may have ripple effects in the local community. When Channel Islands National Marine Sanctuary and the state of California sought to create a network of marine reserves closed to fishing, citizens in the Santa Barbara area and staff from partner agencies shaped the process. Participants analyzed more than 40 alternatives, and between 2002 and 2007, the sanctuary working with the state of California established 11 marine protected areas where no harvests are allowed, and two marine conservation areas where catches are limited. At the time, this was the largest network of marine reserves in the United States. A recent review showed that populations of commercially important species grew both inside and outside the reserves, and demonstrated that reserves had little adverse impact on the local economy.

Citizens play crucial roles in getting sanctuaries designated. For example, Susan Langley, who as a girl was captivated by a photo of an undersea artifact, eventually became Maryland's state underwater marine archaeologist. She was a member of the Monitor National Marine Sanctuary Advisory

Susan Langley, an underwater archaeologist, brings her knowledge to preserving our nation's maritime heritage. She was part of the community-led team that nominated the historic shipwrecks in Maryland's Mallows Bay to become a sanctuary. "The more people who see Mallows Bay, the more excited they become about its shipwrecks. We are protecting this special place."

Council and part of the community-led team that nominated the historic shipwrecks in Maryland's Mallows Bay to become a national marine sanctuary, the first new sanctuary created in the United States since 2000. Langley and others are surveying the new sanctuary to learn more about the massive World War I vessels and other artifacts that date back to the Revolutionary and Civil Wars, and looking for traces of how people used Mallows Bay. "I was just besotted from the beginning," said Langley. "I wanted to see this place become a sanctuary. I take my students here to learn and experience our history. The more people who see it, the more excited they become about Mallows Bay, its shipwrecks, and that we are protecting this special place."

At Mallows Bay, the National Marine Sanctuary Foundation is working with a nonprofit group, called Diving With a Purpose, and students at North Point and Henry E. Lackey High Schools in Charles County, Maryland, to tell the stories of the Ghost Fleet and other wrecks. Diving With a Purpose began in 2005 as a volunteer archaeology program in partnership between members of the National Association of Black Scuba Divers Foundation and the National Park Service. Today the program is a community nonprofit organization that provides education, training, certification, and field experience in maritime archaeology and ocean conservation. Its special focus is the protection, documentation, and interpretation of slave-trade shipwrecks, and the maritime history and culture of African Americans. At North Point and Lackey, students are learning about the science of diving, sea exploration, and national marine sanctuaries.

From American Samoa to Stellwagen Bank, marine sanctuaries rely on a total of 12,300 volunteers who, together, contribute almost 140,000 hours of their time each year. The value of their services to the National Marine Sanctuary System is over $3 million. Sanctuary volunteers inform and inspire students, help collect scientific data on land and on underwater dives, staff visitor centers, and tackle administrative duties. They conduct beach cleanups, track bird populations, and identify whales. The ways to help are as varied as the volunteers. Students and retirees, teachers and techies, veterans and veterinarians, attorneys, rangers, and businesspeople—volunteers represent the diversity and dedication of sanctuary communities.

Because most sanctuaries are large, their scientific research needs are great, yet their staffs are relatively small. Citizen scientists often act as sentinels, alerting the sanctuaries' managers and scientists to potentially important changes and helping to answer some of the sanctuaries' most urgent questions. Armed with water sensors, clipboards, cameras, or dive slates, these volunteers may identify topics that are in need of research, collect and analyze data, interpret results, make new discoveries, and answer complex scientific questions.

Kevin Powers has spent more than 1,000 hours analyzing scientists' data about the seabirds that gather at Stellwagen Bank National Marine Sanctuary off the coast of Massachusetts. For Powers, his volunteer work is a return to his first career as a marine wildlife scientist. In the early 1980s, he conducted surveys of seabirds and whales in the Gulf of Maine, providing crucial information that helped Massachusetts Congressman Gerry Studds's successful effort to nominate Stellwagen Bank for designation as a national marine sanctuary. After his work in field biology, Powers became a computer software engineer, but his love for Stellwagen Bank never left him. He retired from the computer industry in 2012, and the next year he returned to Stellwagen Bank to volunteer with the sanctuary's Seabird Stewards citizen science program. Powers works with the sanctuary scientists who tag and track whales, follow the migrations of great shearwaters that winter off South America and summer at and near Stellwagen, and study a forage fish called sand lance that is an important food source for shearwaters and other birds. His data analysis helps sanctuary managers understand and protect seabirds. He is also a member of the sanctuary's advisory team.

Beach Watch volunteers were out on patrol on the beaches near San Francisco one recent foggy, cold, and damp morning, helping Greater Farallones National Marine Sanctuary monitor seabirds as sentinels of ecosystem health. Birds such as common murres and auklets depend on krill and fish for survival. When natural conditions, such as changes in water temperature, or human-made threats like plastics and toxic chemicals affect prey creatures, they also affect the birds that feed on them. Beach Watch, NOAA's first citizen science monitoring project, was created in 1993 to learn about ocean conditions

Kevin Powers is an internationally recognized seabird researcher whose groundbreaking 1982 research explained the distribution, abundance, and ecological role of marine birds on the continental shelf of the Northwest Atlantic Ocean. His research informed Massachusetts Representative Gerry Studds's successful efforts to nominate Stellwagen Bank for sanctuary designation.

by watching seabirds. Now more than 150 trained citizen scientists conduct 1,300 bimonthly shoreline surveys each year, spanning 175 miles of California coastline. They record live and dead birds and marine mammals, and note human activities that disturb wildlife or violate sanctuary regulations. They report violations, detect oil pollution, and collect evidence of spills that can be used in court cases.

The Channel Islands Naturalist Corps, a joint effort between the sanctuary and Channel Islands National Park, helps collect information on marine mammals and other creatures. In 2011 the program won the Department of the Interior's Take Pride in America award for outstanding federal volunteer work. The next year that award went to Ocean Count, a yearly project in Hawaiian Islands Humpback Whale National Marine Sanctuary. Over the 20-year history of Ocean Count, more than 20,000 volunteers have traveled to more than 60 locations, each of them spending three days in the field.

Some young volunteers find their futures in the sanctuaries. Hannah MacDonald, from Alpena, Michigan, grew up on the shores of Thunder Bay National Marine Sanctuary. As a child she sailed, swam, and dove in the fresh water of the Great Lakes. The shipwrecks lying below the waves fascinated Hannah and captured her imagination. She wondered about the people who had traveled aboard the ships and what their lives and families might have been like. In high school, MacDonald attended the international ocean science and cultural exchange program Ocean for Life at Channel Islands National Marine Sanctuary, volunteered at Thunder Bay, and was named its volunteer of the year in 2014. Two years later she participated in the Big Five Dive, an all-women project to dive among shipwrecks in all five Great Lakes within 24 hours.

LEFT
Hannah MacDonald grew up in Alpena, Michigan, and volunteered at Thunder Bay National Marine Sanctuary as a high school student. "National marine sanctuaries have changed my life and set me on a career path that I am super passionate about, and I'm in love with."

RIGHT
Students in the LiMPETS (Long-term Monitoring Program and Experiential Training for Students) program help survey rocky intertidal zones. LiMPETS is an environmental monitoring and education program for students, educators, and volunteer groups.

By 2019 the recent college graduate was working as an education specialist for the National Marine Sanctuary Foundation in support of NOAA's Office of National Marine Sanctuaries. Like MacDonald, hundreds of young people who volunteer with the sanctuary system learn about the ocean and Great Lakes in sanctuary education programs and participate in NOAA scholarship programs that support studies in marine policy, marine science, or maritime history. They are the future stewards of national marine sanctuaries, the ocean, and Great Lakes.

One of the greatest gifts the sanctuaries offer is enabling people young and old to discover the ocean's secrets. Sanctuaries are living classrooms where people can see, touch, and learn about spectacular marine life and rich maritime history. A citizen science program called LiMPETS—the acronym stands for Long-term Monitoring Program and Experiential Training for Students and plays on the name of a common, cone-shaped marine snail—enlists

1/2" IPS SCH.40 600 PSI@73°F. PVC 1120 AST

FLETCHER

Student teams gather at the headquarters of Thunder Bay National Marine Sanctuary in Alpena, Michigan, to put their engineering skills to the test at the MATE International ROV Competition.

students, educators, and volunteers to monitor the coastal ecosystems of California's national marine sanctuaries. Each year more than 6,000 people collect data at 60 sites spread across 600 miles of coastline. These citizen scientists are the eyes and ears for coastal beaches and rocky shores, detecting changes and possible problems, often before anyone else. The sanctuaries also inspire learning from a distance. Students from inland states follow research expeditions through live streaming of video images from researchers working under distant seas.

At the National Marine Sanctuary of American Samoa, the Summer Sanctuary Science in the Village program works with students to participate in science, education, and community outreach. Throughout the program, students participate in hands-on learning such as marine water testing, beach cleanups, and data collection using scientific sampling methods. The students investigate human impacts on the environment and develop their leadership, critical thinking, and social skills through team activities. According to Chloe Polu, a student in the program, "[It] has truly inspired me to become a marine biologist in the future."

Sanctuaries also encourage students to experiment with new technologies for undersea work. Each year at Thunder Bay National Marine Sanctuary, teams of grade school, high school, and college students gather around a dive training pool, where a series of odd-looking devices, made from metal and plastic piping, scoot across the bottom of the pool. In the sanctuary's annual ROV competition, students design and build underwater vehicles to perform specialized tasks, such as finding a shipwreck, collecting samples of aquatic microorganisms, and tracking invasive species. The competition is staged each year by the Marine Advanced Technology Education (MATE) Center, and many competitions take place in partnership with sanctuaries. Gray's Reef, Olympic Coast, American Samoa, and Monterey Bay national marine sanctuaries also host ROV competitions.

Schools, often the focal point of a community, provide valuable opportunities to build our next generation of ocean guardians. The Ocean Guardian School program, a partnership of NOAA's Office of National Marine Sanctuaries and the National Marine Sanctuary Foundation, provides grants for hands-on watershed stewardship projects to schools in Washington, Oregon, California, Maryland, North Carolina, Texas, Florida, and Hawai'i. The program supports hands-on ocean stewardship projects for kindergarten through 12th grade in five ocean and climate pathways: the 6 R's (rethink, refuse, reduce, reuse, recycle, and rot); marine debris; restoration; schoolyard habitat/garden; and energy and ocean health. The program has served more than 60,000 K–12 students at 134 schools. These small grants provided to schools lead to large impacts: 149,651 kg of trash collected; 726,544 plastic bottles kept out of landfills; 51,656 square meters of nonnative invasive plants removed; 1,600 compost

The Ocean Guardian School program provides grants for hands-on watershed stewardship projects, building the next generation of ocean stewards. Here Ocean Guardian students participate in a beach cleanup with Channel Islands National Marine Sanctuary staff and volunteers.

Letters from Home

From the Pacific Northwest to Polynesia, young people growing up in traditional cultures, including those alongside the national marine sanctuaries, face big challenges from the twin forces of climate change and culture change. To help them prepare to protect their waters and the wider ocean, NOAA's Pacific Northwest Bay Watershed Education and Training Program started an environmental and cultural exchange program called Ecosystem Pen Pals in 2016. Taking part were 190 high school students near national marine sanctuaries in the Hawaiian Islands, American Samoa, and Washington State. They documented their own ecosystems, producing field guides, videos, and posters. They sent "ecosystem suitcases" filled with natural and cultural artifacts to students involved with the other participating sanctuaries, and exchanged letters about global issues such as food sovereignty; the right to healthy, culturally appropriate, sustainably produced food; ocean acidification; and climate change.

On Earth Day 2016 the Suquamish Tribe, Olympic Coast National Marine Sanctuary, and the nonprofit group EarthEcho International hosted the Indigenous Youth Summit on Climate Change and Ocean Change. Students from each of the participating schools gave presentations on the effects of global change on their traditional cultures and natural resources. The event was a celebration of Earth as a blue planet and a conversation about the urgent need to work together to solve ocean problems around the Pacific Rim.

and recycle bins installed; and 36,730 reusable bags and bottles distributed to replace single-use plastic items.

Gault Elementary School, an Ocean Guardian School located a mile's walk from the shores of the Monterey Bay National Marine Sanctuary, offers another example of how youth are leading the way. With help from local partners and the community, the students of Gault set out to restore the coastal dunes at their neighborhood Seabright Beach. Over many years, these young ocean guardians have removed over 4,000 square meters of invasive plants and planted 3,500 square meters of more than 12,000 native plants that they had helped propagate in their school greenhouse. As Gault students grew more invested in the project, they began to talk about their work with their families and friends and proudly reported their progress to their city council. Five years later, they experienced firsthand how their stewardship efforts had helped rebuild their living shoreline: they witnessed the return of two endangered species—the snowy plover and the burrowing owl—to their beloved beach.

Each year the sanctuary system's programs reach over 25,000 students, 2,000 educators, and 58,000 lifelong learners. Many others learn from sanctuary activities, lesson plans, visitor centers, exhibits, and videos. Ocean

By eating lionfish, you can help reef ecosystems in the southeastern United States, Caribbean Sea, and Gulf of Mexico. Lionfish are highly invasive species in these areas, and due to their voracious appetites and lack of natural predators, they're rapidly edging out the native species that reefs need to remain healthy.

literacy—understanding the human impact on the ocean and its impact on us—is at the heart of education programs that inspire conservation, spark the imagination of the next generation, and encourage people of all backgrounds to connect to these special places. Sanctuaries offer a chance to dive into a world of beauty and abundance.

Each year the sanctuary system benefits from NOAA scholarships, internships, and fellowships. The Dr. Nancy Foster Scholarship Program seeks to bring more diversity to the ranks of marine scientists. Sixty-five graduate students, most of them women, received grants for college expenses and research projects in sanctuaries from Channel Islands National Marine Sanctuary to Stellwagen Bank. Lindsay Marks, a graduate student from California who participated in the scholarship program, is fighting back against devil weed, an Asian species that is invading waters along the West Coast. Devil weed blankets the seafloor with a dense, low canopy, crowding out native giant kelp as well as the fish and other wildlife that typically live in kelp forests. One way to control the invasion is underwater "weeding," or removing the plants with a vacuum pump. This approach can reduce the next generation of devil weed by 40 percent, giving native species a breather from the onslaught.

Reef Rescuers

Coral is a living organism, a colony of marine animals that extract minerals from seawater and use them to form their skeletons, made primarily of calcium and carbon. Each coral animal lives in partnership with a particular type of single-celled marine algae. The algae's photosynthesis helps nourish the coral animal in clear, tropical waters, which are generally low in nutrients, while the coral's skeleton protects the algae from being eaten. These limestone cities teem with life. Climate change, coral diseases, pollution, and extreme heat easily disrupt this symbiotic relationship, causing corals to expel the algae in a reaction called coral bleaching that can cause large swaths of the reef to die. Ship groundings, derelict fishing gear, or carelessly thrown boat anchors also injure corals, which may need years or decades to recover.

A coral reef nursery in Florida Keys National Marine Sanctuary helps to regrow coral fragments.

The coral reefs of the Florida Keys, like most Caribbean coral reefs, have suffered steady declines since the 1970s. Nearly 90 percent of the live corals that once dominated the reefs have been lost. There is no single cause for their decline. Climate change has at times made the Keys' waters too hot for corals, which can tolerate only a narrow range of temperatures; as a result, there have been extensive coral die-offs. Boat groundings, overfishing, pollution, runoff from coastal development, and other human impacts have caused destruction and disease. In 2019 NOAA and partners launched an unprecedented, decades-long approach to restore seven iconic coral reef sites in Florida Keys National Marine Sanctuary. The selected reef sites represent a diversity of habitats, show a high probability for successful restoration, and support the quality of life and economic health of local communities. The effort represents one of the largest investments ever undertaken in coral restoration. Scientists are slowly growing new colonies, initially using living fragments rescued from seawall construction projects or salvaged from boat groundings and from colonies that show signs of resistance to bleaching and other stresses. The coral fragments are strung from underwater nursery "trees"—structures with many arms where growing coral pieces hang suspended in the water column like ornaments on an undersea Christmas tree. Once the pieces grow large enough, they are either fragmented to grow more corals, or divers will transplant them to the reef tract. These nursery-raised corals are spawning in synchronicity with the phases of the late summer moon, just as their wild-growing kin do. This gives the experts hope that the coral transplants will survive, reproduce, and help the reef rebound.

Corals are also found in the deep seas, where they are just as vulnerable to impacts from fishing, mining, and pollution. In California's Monterey Bay National Marine Sanctuary, scientists are collaborating with the Monterey Bay Aquarium Research Institute on new techniques for deep-sea coral restoration. This is the first time researchers are attempting to develop and test restoration methods for deep-sea coral species in the Pacific Ocean. Exploring the ocean depths used to be difficult, but now precision navigation technologies enable scientists to pinpoint the exact locations where they have transplanted deep-sea corals and to return later to check the test beds.

The Ernest F. Hollings Undergraduate Scholarship, named after the late US senator and champion of ocean conservation, sponsors two years of study in a field related to NOAA's work and an internship with a NOAA program. The Education Partnership Program with Minority Serving Institutions also provides NOAA internships and two-year scholarships for undergraduates majoring in science, technology, engineering, or math. The Sea Grant Knauss Fellowship, named for former NOAA administrator John A. Knauss, offers exceptional graduate students one-year paid fellowships with federal agencies in the Washington, DC, area.

Stewardship of our ocean and Great Lakes can take many forms, including engaging chefs to serve only sustainable seafood or prepare delicious meals with invasive species. Most sanctuaries contain some invasive species. These invasive species can harm the living communities where they are introduced by becoming too abundant or eating too much of the native marine population. They can reduce the diversity and abundance of native species, upset the stability of invaded ecosystems, damage archaeological resources, and interfere with farming, aquaculture, and recreation. Invasive species can even cause local extinctions of native species. One example is the Indo-Pacific lionfish, an aggressive feeder that is rapidly spreading in the Atlantic, the Gulf of Mexico, and the Caribbean, exterminating many native fish populations. Three sanctuaries are working on eradicating invasive lionfish by encouraging people to catch and eat them. The Florida Keys and Flower Garden Banks national marine sanctuaries host lionfish tournaments, where fishers remove as many of the pesky creatures as they can. Some local chefs feature lionfish on their menus. Each year, Gray's Reef National Marine Sanctuary Foundation, a chapter of the National Marine Sanctuary Foundation, makes lionfish the centerpiece of a dinner called "A Fishy Affair: Malicious but Delicious."

A person does not need to live on the coast to protect the ocean and Great Lakes. Construction materials, lost fishing gear, and an endless stream of every imaginable kind of cast-off plastic product, from cigarette lighters to water bottles, litter our waters. Everyone contributes to it. Our debris is carried by winds, rivers, and currents to every part of the ocean. Marine debris threatens habitats and wildlife, interferes with navigation, and in some cases injures human health. We all have a role to play in reducing marine debris and plastic pollution.

Turtles, marine mammals, birds, and bottom-dwelling creatures mistake plastic for food, eat it, and starve to death. Plastic ingested by an animal takes up room in its stomach and can prevent digestion of food, causing a slow and painful death. When people release helium balloons to celebrate a special occasion, they may not realize that these balloons eventually deflate and can end up in the water, where they frequently choke turtles and seabirds. Turtles, dolphins, and whales, among others, get entangled in nets or plastic. When this

Whales and other marine mammals are not the only sea creatures to fall victim to entanglement. Here a sea turtle is freed from debris.

happens, they can drown, starve, or suffer wounds and infections that can fatally weaken them. One of the most insidious problems is ghost fishing: lost or discarded fishing gear that continues to drift, catching and killing fish for years.

Almost every sanctuary must deal with marine debris. Papahānaumokuākea Marine National Monument encompasses some of the most isolated islands on Earth, and in some ways the area seems untouched by humans. Yet ocean currents still carry an estimated 50 tons of plastic, nets, and other junk to these islands and surrounding reefs every year. To collect debris from these remote islands is a massive undertaking, requiring special trash-removal voyages by skilled divers.

At many sanctuaries, communities get involved in the cleanups, which make a dent in the problem and raise awareness about how to keep trash out of the ocean in the first place by disposing of plastics appropriately. In September 2017, Hurricane Irma struck the Florida Keys, damaging homes, sinking vessels, and scattering debris into the marine environment. The lost and damaged fishing gear from the hurricane, compounded with chronically lost gear and other forms of marine debris, poses a high risk for damaging

Marine debris is an increasing problem in sanctuaries. Entanglement of marine mammals results in drowning, starvation, physical trauma, and systemic infections, or increases susceptibility to other threats such as ship strikes. The North Pacific stock of humpback whales that migrates to Hawai'i each winter is significantly threatened.

critical habitats such as reefs and seagrass beds and can entangle and harm corals, sponges, dolphins, manatees, and sea turtles. In May 2018, the Office of National Marine Sanctuaries launched Goal: Clean Seas Florida Keys to remove underwater marine debris from Florida Keys National Marine Sanctuary and educate the public about its role in marine debris prevention. Goal: Clean Seas Florida Keys partners with sanctuary-recognized Blue Star Dive Operators to educate dive professionals and recreational divers on best practices for removal of marine debris. In the first year of the program, National Marine Sanctuary Foundation–funded divers conducted 49 cleanup trips, engaged 450 volunteer divers, and spent nearly 900 hours underwater removing 78 intact lobster traps, hundreds of pieces of lobster trap debris, 16,369 feet of line, and 14,693 pounds of debris from Florida Keys National Marine Sanctuary. Habitats in the Florida Keys benefit greatly from the removal of marine debris, and participating dive shops help with the stewardship of this special place.

Ocean guardians are often leading the way toward banning plastic straws and balloon releases, and reducing the use of bottled water, plastic packaging,

Fishing nets and plastics are major problems facing our ocean. Here nets are being removed from Papahānaumokuākea Marine National Monument.

HONDA
HONDA
30
3

The Hawaiian monk seal is one of the most endangered seal species in the world. The population overall has been declining for over six decades, and current numbers are only about one-third of historic population levels. Beaches that are popular for human recreation are also used by monk seals for "hauling out" and molting, and some female monk seals are also pupping on popular recreational beaches.

and plastic bags. Since 2014 J. C. Parks Elementary School students in Indian Head, Maryland, have worked through their Ocean Guardian School program to reduce waste on their campus and in the community, especially single-use plastic waste. Their efforts have included improving the school's recycling program with the addition of 43 recycling bins; distributing 750 reusable bottles; and installing a hydration station where students have prevented over 100,000 water bottles from going to landfills. As their stewardship projects grew on campus, students became more involved in raising awareness in their community about how actions on land affect the health of marine habitats. In 2017 the J. C. Parks Green Team led the school's "Skip the Straw" campaign, which received a formal citation from Clark County commissioners. After the J. C. Parks Ocean Guardians wrote letters, attended public meetings, and delivered presentations, in October 2018 the commissioners passed Bill 2018-07, banning plastic straws and stirrers from all Charles County restaurants and businesses.

Entanglement in fishing lines and nets and marine debris poses particular threats to cetaceans worldwide. NOAA fields special teams of highly trained experts and volunteers who risk their lives to save entangled whales. The task requires in-depth knowledge of whale behavior and biology. Even with training, experience, and an abundance of caution, it can be extremely dangerous. Rescuers never enter the water with an entangled whale. Instead, they grab hold of the tangled fishing lines using a grappling hook and attach buoys to the fishing lines. These keep the whale at the surface and slow it down enough for the team to catch up to it in a small, inflatable boat. Once they are close enough, they use a custom-designed cutting tool on a long pole to cut away the snarled gear without injuring the whale. Once the whale is free, the team uses the grappling hook to remove the debris from the water so it can't entrap another animal. Teams also try to identify the source of entangling nets or lines in hopes of reducing the threat in the future.

Humankind is profoundly affecting ocean life. That makes our stewardship of the oceans more important than ever. Battalions of volunteers, researchers, and advocates are sorely needed at a time when human activities are putting our ocean and Great Lakes at greater risk. As forest scientist Baba Dioum told an assembly of conservationists in 1968, "In the end we will conserve only what we love, we will love only what we understand, and we will understand only what we are taught." National marine sanctuaries and monuments are places where we can learn more about our precious seaways and engage with communities in caring for them. They increase our understanding of the Earth and the magnificent species and seascapes that we share it with, as well as how human actions are putting them and our own livelihoods increasingly at risk. And they invite us to become better stewards, conserving our ocean and Great Lakes for both current and future generations. ○

SUSTAINING OUR OCEANS AND GREAT LAKES

BY KRISTEN SARRI, PRESIDENT AND CEO, NATIONAL MARINE SANCTUARY FOUNDATION

We need to respect the oceans and take care of them as if our lives depended on it. Because they do.

SYLVIA EARLY, TRUSTEE EMERITA,
NATIONAL MARINE SANCTUARY FOUNDATION

THE TERM *SANCTUARY* HAS MANY MEANINGS: a place of refuge, a holy place, or a nature reserve. Each of these meanings applies to our national marine sanctuaries. They offer refuge to the largest of the ocean's creatures—the blue whale—and to some of its smallest—the tiny, shrimplike krill that, at less than an inch long, are essential food for the blue whale and countless other animals. They are the hallowed resting places for men and women who lost their lives in commerce and combat. And they protect extraordinary seascapes that are home to endangered, threatened, and rare species.

When I visit one of the national marine sanctuaries, the smell of the salty sea air, the whoosh of the waves on the shore, and the cold shock of the water when it rushes through my toes transport me back to some of my favorite memories: growing up in Michigan, canoeing with my dad; getting spun head over heels in the waves on family beach vacations; and staring for hours at the variety of life and activity within tide pools. I am thankful that in 1972, 100 years after the establishment of our first national park, the National Marine Sanctuary Act gave us the means to permanently protect the most nationally significant places in our ocean and Great Lakes.

Today the National Marine Sanctuary System encompasses 14 national marine sanctuaries and Papahānaumokuākea and Rose Atoll marine national monuments. Together these protected waters cover over 600,000 square miles. The wonders that lie beneath the waves are hard for most people to see. The pages of this book beautifully reveal these wonders, from the large Porites coral found in American Samoa's Valley of Giants across the Pacific to California's giant kelp forest, and across the continent to the murky green waters of the Atlantic and the "live bottom" of Gray's Reef. This book is an invitation to discover the wonders of our ocean and Great Lakes—wonders that belong to all of

A snorkeler discovers the wonders of Hawaii's reefs. Coral reefs are the most biologically diverse marine ecosystems, yet they only cover a small portion of the ocean floor.

A pod of humpback whales at Stellwagen Bank National Marine Sanctuary is bubble feeding on prey fish. A group of whales will round up the fish into a tight circle and blow bubbles; the net of bubbles surrounding the fish keeps them from escaping.

Orcas, commonly called killer whales, are one of the world's most powerful predators. Orcas are social and use a variety of sounds to communicate with each other. Each pod has distinctive sounds that its members will recognize.

us, whether we live along the coast or in the heartland—and, through the eyes of explorers, to learn of the shipwrecks that document our history and provide a further connection to the seas.

The National Oceanic and Atmospheric Administration's Office of National Marine Sanctuaries holds these special places in trust for current and future generations. On shore and at sea, scientists, managers, and educators working in national marine sanctuaries help us understand more about our ocean and Great Lakes. Listening to soundscapes in the water and observing both close up and remotely, researchers study animals and gather information that aids in their conservation. Through long-term monitoring, they help to detect changes in the environment and add to our knowledge of how climate change, sea-level rise, and ocean acidification are altering our waters and the wildlife that live in them. Exploration and mapping expeditions in sanctuaries hold enormous potential for the discovery of new species, including some found nowhere else on Earth. Sanctuaries and monuments are ideal places for scientists, students, and the public to study our ocean and Great Lakes in never-before-possible ways through such innovations as Remotely Operated

Vehicles, telepresence, and underwater imaging. These technologies are making virtual visits a reality and are leading to better understanding of our protected waters to ensure their long-term sustainability. Marine archaeologists working in sanctuaries and monuments give us insights into our country's maritime and cultural heritage by exploring and interpreting prehistoric sites, shipwrecks, and naval battlefields. Through their work, we can better understand how our ocean and Great Lakes shaped human settlements, immigration, economies, and cultures, and can gain a greater appreciation for why we need to protect the ocean and Great Lakes.

Founded in 2000 by America's most influential ocean conservation leaders, the National Marine Sanctuary Foundation works with communities, and in close partnership with NOAA, to protect our waters and to preserve America's maritime heritage and cultural history. Volunteers, citizen scientists, and local businesses are key to the stewardship of our sanctuaries and monuments. They help monitor their health, conduct research, restore fragile ecosystems, remove marine debris, and educate students and the public. Through on-the-water conservation efforts, scientific research, and education, the National Marine Sanctuary Foundation invites people with a stake in the health of their planet to protect their place in it. We work to inspire new guardians for our ocean and Great Lakes, and to demonstrate how the sanctuaries can serve as a model for protecting marine ecosystems across the globe.

Renowned marine ecologist and former NOAA Administrator Jane Lubchenco wrote about three narratives of our attitudes toward the ocean. The first narrative, which dominated most of human history with the ocean: it is too bountiful and big to fail. The second narrative: humans are depleting resources at an alarming rate and disrupting the ocean; it is too big to fix. The third narrative: the ocean is central to our future, and it is too big and important to ignore. Our communities and economies rely on a healthy ocean and Great Lakes. Throughout the world, we are losing species diversity, with harmful consequences for humans, economies, and the environment. But there are solutions, if we work together as stewards of nature. Increasingly, scientists, managers, and governments are calling for countries to work together across national boundaries, to protect more of nature and to reduce the loss of diversity, slow climate change, and increase our ability to adapt to it. Our current sanctuaries and monuments are an important part of this global effort. But there is more we can do.

With jurisdiction over almost 4.5 million square miles of ocean—an area 23 percent larger than our nation's landmass—the United States plays a critical role in protecting our global ocean. These waters span a multitude of ecosystems, from the icy-cold gray waters of the Arctic to the tropical blue waters of the Caribbean Sea. By designating new national marine sanctuaries and marine national monuments in areas that represent different marine and freshwater ecosystems, expanding existing ones, and strengthening protections for biodiversity, we can do our part in the global effort to conserve nature and maintain a healthy ocean and Great Lakes.

In 2022 the National Marine Sanctuary Act will celebrate its 50th anniversary. Looking ahead to the next 50 years, the National Marine Sanctuary Foundation will continue to work with NOAA and communities across the country to expand our National Marine Sanctuary System. For citizens and communities interested in creating a new national marine sanctuary, there are two different approaches to pursue. Citizens can advocate to their members of Congress to pass

There is nothing like the sheer joy of jumping on rocks as waves crash.

legislation to create a national marine sanctuary—legislation that is later signed into law by the president. Such laws established Florida Keys, Stellwagen Bank, Monterey Bay, and Hawaiian Islands Humpback Whale national marine sanctuaries. Communities can nominate their most treasured places in our marine and Great Lakes waters for consideration as national marine sanctuaries by submitting a proposal through the sanctuary nomination process. NOAA's Office of National Marine Sanctuaries reviews these community-based nominations to ensure they have diverse support and meet criteria of national significance. After NOAA accepts a candidate site, then it may be considered for potential designation.

Sanctuary designation is a separate public process that is highly public and participatory, and often takes several years to complete. It emphasizes community participation and engagement to ensure that the national marine sanctuary considers the needs of interested communities. The designation process includes public meetings, comment periods, and consultation to inform NOAA's development of the management plans and regulations. Nominations can remain in the inventory for up to five years before they are reviewed and updated. In 2019 Mallows Bay–Potomac River was the first new sanctuary designated in almost 20 years. As of this writing, two nominations, Lake Michigan (Wisconsin) and Lake Ontario (New York), are in the designation process. And, hopefully, they will become new sanctuaries in the next year. NOAA is reviewing one nomination for listing in the inventory: Shipwreck Coast in Lake Superior (Michigan). There are five nominations in the inventory: Hudson Canyon (New York); St. George Unangan Heritage (Alaska); Lake Erie Quadrangle (Pennsylvania); Chumash Heritage (California); and Mariana Trench (Northern Mariana Islands). With the support of passionate advocates in these and other communities across the country, we are confident that this is just the beginning of a new era for ocean and Great Lakes conservation in the United States.

We are all stewards of our ocean and Great Lakes—for the species that depend on them and for future generations. Through sanctuaries and monuments, we can work collectively to conserve these resources for all Americans to enjoy. We need to be their voice in the halls of Congress, state legislatures, city halls, and corporate board rooms advocating for greater protection and sustainable business practices that help them thrive for future generations. We invite you to join us in our efforts to advocate for more protected areas in our ocean and Great Lakes, especially those that strongly safeguard natural and cultural resources.

Each year in June, the National Marine Sanctuary Foundation hosts Capitol Hill Ocean Week (affectionately known as CHOW) to bring together scientists, conservation organizations, business leaders, and government officials to discuss pressing policy, conservation, and science issues facing our ocean and Great Lakes and innovative solutions to address them. CHOW is open to the public and welcomes all who want to join in the dialogue and take action to help conserve our ocean and Great Lakes.

Supporting our marine sanctuaries and monuments connects us to our communities, our country, and our world. They fill those who know them with wonder. They beckon us to explore beyond our horizons. Each one protects a habitat unlike any other on Earth. Marine sanctuaries and monuments are places where we can conserve our waters, for the good of the world and everything in it. ○

DISCOVER WONDER

Profiles of the National Marine Sanctuaries and Monuments

National Marine Sanctuary of American Samoa

Designated: *April 29, 1986*
Size: *13,581 square miles*
Location: *American Samoa*

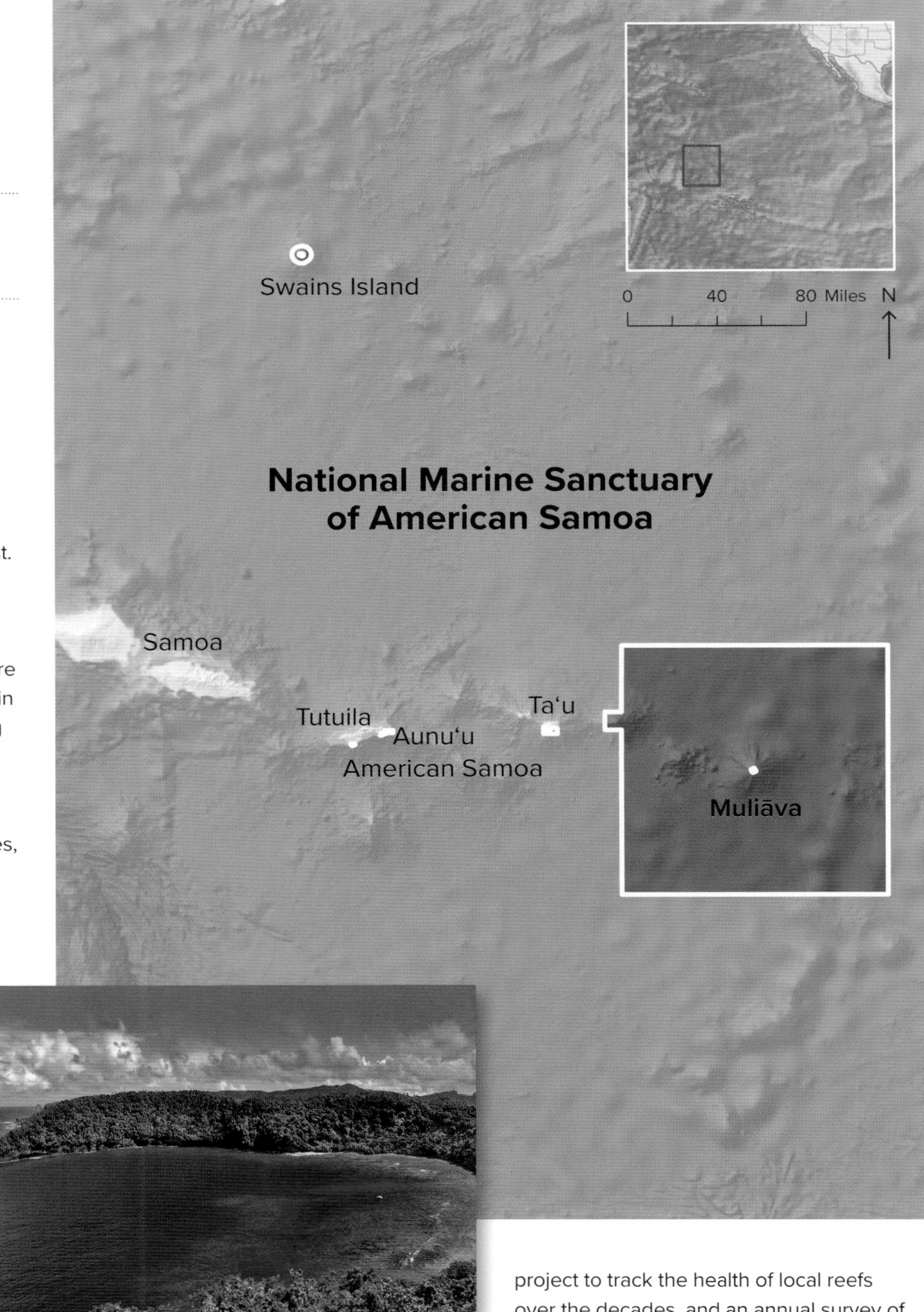

Fagatele Bay protects a diverse coral ecosystem.

In the waters between Hawai'i and New Zealand, and in the heart of Polynesian culture, lies the National Marine Sanctuary of American Samoa. Once the smallest sanctuary, centered on a .25-square-mile reef, the sanctuary added large swaths of near-shore coral reefs and offshore open ocean waters in 2012, and is now the largest.

American Samoa's tropical reef is a dazzling tapestry of color and diversity, with more than 150 different coral species. Massive Porites coral heads at Ta'u Island are the oldest and largest corals of their genus in the world; one of them, Fale Bommie, or Big Momma, is over 500 years old and 21 feet high. It's common to see sea turtles gliding gracefully through an undersea forest of coral. With 1,400 marine invertebrate species, 950 species of brilliantly colored fish, more than a dozen kinds of marine mammals, all in water as clear as liquid sunlight, the underwater world transforms moment by moment, just like the turn of a kaleidoscope.

In the water and on land, natural and cultural artifacts are nestled in places with special meaning in the islands' traditions, encompassing 3,000 years of American Samoan history and community life. Resting on the white sand of the ocean floor sit wrecks of whalers and other lost ships. World War II relics like coastal pillboxes and gun emplacements lie scattered across beaches. These artifacts are reminders of a shared history; they serve as gateways into the past.

The sanctuary carries out its work in a way that honors Fa'a-Samoa, the Samoan Way. For instance, the ancient concept of *tapu* restricts the use of overly stressed areas in order to protect them and give their resources time to replenish. In keeping with the concept of tapu, the sanctuary continues the tradition of protecting and preserving the cultural treasures and natural resources of this land and ocean.

The sanctuary is also a center for research and monitoring, including a comprehensive project to track the health of local reefs over the decades, and an annual survey of humpback whales. To get the public involved in conservation, research, and cultural activities, the sanctuary holds a marine science camp and community events that tie in to traditional Samoan culture. Visitors can explore the sanctuary at the Tauese P. F. Sunia Ocean Center. Opportunities to explore include snorkeling, diving, wildlife viewing, and fishing or hiking in the nearby National Park of American Samoa.

Channel Islands National Marine Sanctuary

Designated: *September 22, 1980*
Size: *1,470 square miles*
Location: *Southern California*

Off the coast of Southern California lies Channel Islands National Marine Sanctuary, which encompasses ocean waters surrounding Anacapa, Santa Cruz, Santa Rosa, San Miguel, and Santa Barbara islands. Near-shore waters within the sanctuary are co-managed with Channel Islands National Park and the state of California.

The ancestral home of the Chumash people, these islands have a rich maritime history and lasting cultural significance. For millennia before European contact, the Chumash crossed between the islands and mainland in canoes known as *tomol*s. Following the arrival of Europeans, the region became a busy route for trade, passenger and military ships sailing up and down the coast. The strong currents caused many shipwrecks—and the islands' dense fog caused many plane crashes as well. Most remain under water, undiscovered.

Santa Barbara Harbor is a gateway to the Channel Islands.

The sanctuary protects exceptional biodiversity, productive ecosystems, sensitive species and habitats, and prized fishing grounds. Deep-sea corals and sponges are nurseries for commercially valuable fish. Undersea forests of giant kelp shelter brightly colored invertebrates. Sandy beaches host colonies of sea lions and seals, like the Northern elephant seals who breed on San Miguel Island in January and February. Rocky rises draw seabirds to breed, nest, and feed. Twenty-seven species of marine mammals, from blue whales to porpoises, pass through sanctuary waters, breaching, surfing waves, and delighting observers onshore and on whale-watching vessels.

The Channel Islands Naturalist Corps volunteers provide education programming aboard local whale-watch vessels. Outdoor recreational opportunities for visitors to the sanctuary include scuba diving, kayaking, boating, and tide pooling, and of course, spectacular wildlife watching.

Cordell Bank boasts a seascape like no other.

Cordell Bank National Marine Sanctuary

Designated: *May 24, 1989*
Size: *1,286 square miles*
Location: *Northern California*

In 1869 the intrepid marine surveyor Edward Cordell explored the offshore rise that bears his name. Today Cordell Bank National Marine Sanctuary is a lush feeding ground for many marine mammals and seabirds.

The 1,286-square-mile sanctuary includes the bank, which is 22 miles off the coast north of San Francisco, the undersea Bodega Canyon, and granite and rock reefs. The bank itself is a 4½-by-9½-mile rocky undersea structure that sits at the edge of the continental shelf, where every March to July the southward-flowing California Current creates a nutrient-rich upwelling that supports an abundance of sea life.

Sponges, sea stars, sea cucumbers, and moon jellies thrive in the sanctuary's water, which also attracts feeding rockfish, tuna, mackerels, and, in summer, ocean sunfish. The seafood banquet draws large marine mammals like humpback whales, Pacific white-sided dolphins, Dall's porpoises, and California sea lions. Studies show that salmon and sharks also forage here. An array of seabirds, including five species of albatross, flock to the sanctuary during the spring and summer upwelling season. Some, like Cassin's auklets, stick around longer, feeding year-round on krill.

Cordell Bank's diversity makes the sanctuary a prime marine habitat for deep-sea scientific studies and exploration. Researchers focus on characterizing the species and habitats of the bank and surrounding areas. Since the bank lies a distance from shore, weather and water can make expeditions challenging, but scientists use side-scan sonar and remotely operated vehicles to study the sanctuary. They have discovered a new species of coral in the sanctuary, and have deployed an acoustic buoy to record underwater sound, as part of a research effort to document the soundscape of natural and human-made noises, and better understand its importance to ocean creatures.

Because the bank is not easy to visit, sanctuary educators bring the wonders of Cordell Bank to people onshore. A radio program, *Ocean Currents*, airs on local station KWMR once a month and is available as a podcast. The Oakland Museum of California, Bodega Marine Laboratory, and Point Reyes National Seashore Visitor Center host exhibits about the sanctuary.

Florida Keys National Marine Sanctuary

Designated: *November 16, 1990*
Size: *3,840 square miles*
Location: *Florida Keys*

Florida Keys National Marine Sanctuary protects the turquoise waters surrounding 1,700 miles of islands, creating a mosaic of marine habitats where around 6,000 species of plants and animals thrive. Herons, egrets, and rare mangrove cuckoos feed on the tidal flats, seagrass beds, and along the mangrove-forested shoreline, while hawks, eagles, and ospreys swoop down to fish in the channels. Manatees drift through seagrass beds and sheltered coves, bottlenose dolphins and sea turtles glide throughout the sanctuary's waters, and tropical fish swarm around vibrant coral reefs.

The sanctuary protects the only barrier coral reef in the continental United States. Stretching close to 150 miles, the reef is home to more than 45 species of stony corals and 35 soft corals. Seven of the coral species are listed as threatened under the Endangered Species Act. The sanctuary also includes one of the world's largest seagrass beds. Seagrass stabilizes the sea bottom, helps to maintain water clarity, and provides critical habitat for fish and invertebrates and foraging areas for turtles, manatees, parrotfish, and sea urchins.

Colorful angelfish brighten the waters of the Florida Keys.

Over 2,000 vessels lie along the coral reefs and buried in the sandy shallows of the Florida Keys. Divers and snorkelers can explore the skeletal remains of nine shipwrecks, the oldest from 1733, on the underwater Florida Keys Shipwreck Trail. The sanctuary is also a paradise for kayaking, paddle boarding, sailing, and fishing. Those who prefer to stay on land can take in the interactive exhibits at the Florida Keys Eco-Discovery Center and can enjoy the Keys' abundant wildlife, especially in spring and fall when the island chain is a way station for migratory songbirds traveling across the Caribbean Sea between North and South America.

Flower Garden Banks National Marine Sanctuary

Designated: *January 17, 1992*
Size: *56 square miles*
Location: *Gulf of Mexico off Texas and Louisiana*

In the warm waters of the Gulf of Mexico exists a splendor of marine communities. Flower Garden Banks, the only marine sanctuary in the gulf, was first discovered by snapper and grouper fishers in the early 1900s. The sanctuary gets its name from the sponges, plants, and diversity of marine life the fishers experienced upon discovering the coral reefs. These living communities grow and thrive atop three underwater mountains, or banks, which range from 55 to almost 500 feet deep.

Flower Garden Banks National Marine Sanctuary is brimming with activity, from quick-swimming snapper and jacks, to graceful manta rays and sea turtles, to colorful reef fish and invertebrates. Of special note is the golden smooth trunkfish, a variant of a common Caribbean fish that is only found here; and the Mardi Gras wrasse, a fish just discovered and described in the past 15 years. Also stunning is the mass coral spawning that occurs at night every August, in synchrony with the late summer moon. Against the velvety black water, luminous clouds of coral eggs and sperm look like an undersea Milky Way.

An oceanic manta ray glides through Flower Garden Banks.

In 1992 Flower Garden Banks National Marine Sanctuary was designated to protect the habitat and marine life of East and West Flower Garden Banks. Stetson Bank was added in 1996. Now, the sanctuary is considering an expansion of 150 square miles to incorporate other reefs and banks that include a variety of undersea bottom features with similar biology. This expansion will protect even more wondrous creatures and flourishing habitats so future generations may enjoy and explore the Gulf of Mexico's vibrant world.

Although the sanctuary lies 70 to 115 miles from the coastline, diving and fishing charters bring adventurous visitors to the banks for offshore experiences they won't soon forget.

Gray's Reef National Marine Sanctuary

Designated: *January 16, 1981*
Size: *22 square miles*
Location: *Off the Georgia Coast*

Gray's Reef is home to colorful coral and sponge forests.

In 1961 biological collector and curator Milton "Sam" Gray was surveying the ocean floor 20 miles off Sapelo Island, Georgia, when he uncovered a hidden treasure—a hard-bottom reef bubbling with life. Today this ecosystem encompasses the 22 square miles of protected marine area known as Gray's Reef National Marine Sanctuary. The sanctuary is currently the only protected natural reef area on the continental shelf off the Georgia coast.

Unlike tropical reefs formed by hard corals, Gray's Reef is a conglomeration of sediment, like shell fragments and sand, that arrived by coastal rivers and currents. This unique reef structure creates a "live bottom habitat"—a hard, rocky seafloor. "Live bottom" refers to hard or rocky seafloor that typically supports high numbers of large invertebrates such as sponges, corals, and sea squirts. These spineless creatures thrive in rocky areas, as many are able to attach themselves more firmly to the hard substrate, as compared to sandy or muddy "soft" bottom habitats.

The scattered rocky outcroppings and ledges of Gray's Reef provide homes for an abundance of marine life. The reef attracts more than 200 species of fish, including black sea bass, snappers, groupers, and mackerels. With a wealth of sport fish, Gray's Reef is a popular fishing location, especially with anglers looking to catch the big ones.

The diving here is spectacular as well. In addition to the variety of fish, the reef is home to diverse organisms that form a dense, vibrant carpet of living creatures in every color of the rainbow. Since the reef fauna changes seasonally, no two dives are ever alike.

The sanctuary conserves the reef's marine life, from vibrant sea stars and nudibranchs to schools of fish, gliding loggerhead sea turtles, and hammerhead sharks. A portion of the sanctuary is dedicated to research, where scientists study fish and invertebrate biology and behavior, characterize marine habitats, and monitor conditions to better understand how natural and human-caused changes are affecting the undersea environment.

Greater Farallones National Marine Sanctuary

Designated: *January 16, 1981*
Size: *3,295 square miles*
Location: *Central California Coast*

The California Current sweeps down the West Coast of North America from Alaska to Mexico. Nutrients carried by the cold waters of the current make Greater Farallones National Marine Sanctuary one of the most diverse and bountiful marine environments in the world.

Reaching from San Francisco north to the scenic Redwood Coast, the sanctuary gets its name from the Farallon Islands, a set of foreboding, rocky islands off San Francisco. (The Spanish word *farallon* means "cliff" or "outcrop.") The islands support an extraordinarily diverse marine mosaic of kelp forests, deep-sea coral communities, rocky intertidal zones, sandy beaches, and *esteros*, or marshy estuaries.

The sanctuary is the feeding and breeding grounds and migratory pathway for at least 25 endangered and threatened species, 37 marine mammal species, and one of the most significant white shark populations on the planet. It is also home to over a quarter million seabirds—the largest concentration of breeding seabirds in the continental United States. Auklets and puffins are among the 13 seabird species that nest on the islands. Wading birds feed in the esteros.

Hundreds of marine invertebrate species live in its waters, including deep-sea corals, sponges, shrimp, crabs, and other mollusks and crustaceans. Nearly 400 species of fish use every part of the sanctuary to feed, migrate, rest, and reproduce. Wetlands in the sanctuary are designated as globally important under the Ramsar Convention, an international treaty to protect wetlands.

The sanctuary protects submerged prehistoric remains, sacred tribal areas, historic ship and aircraft wrecks, and the remnants of docks, wharves, and piers that once served logging and fishing. It offers an array of outdoor activities and educational opportunities within reach of the San Francisco Bay area, with options ranging from water activities like whale watching, boating, kayaking, and surfing to shore adventures like hiking, birdwatching, and tide pooling. The sanctuary also supports a Marine Explorers day camp and popular events like Beach Watch, and conducts research, such as the White Shark Stewardship Project, that welcome the participation of citizen scientists.

Point Arena Lighthouse is on the sanctuary's coast.

The beaches in Hawai'i are perfect for swimming.

Hawaiian Islands Humpback Whale National Marine Sanctuary

Designated: *November 4, 1992*
Size: *1,366 square miles*
Location: *Hawai'i*

Humpback whales (*koholā* in Hawaiian) are *'aumākua*, the sacred ancestors and spirits of the Hawaiian people. The shallow, warm waters that swirl around the main Hawaiian Islands are the only places in the United States where these gentle giants return each year to breed, give birth, and care for newborn calves.

Hawaiian Islands Humpback Whale National Marine Sanctuary was created in 1992 to protect about 10,000 humpback whales and their habitat, and to recognize the importance of Native Hawaiian cultural traditions and their relationship to the long-term health of the ocean. Within the sanctuary are the remains of traditional Hawaiian aquaculture ponds where fish were raised for food, and traces of fishing tools and other artifacts relating to coastal settlements.

The sanctuary encompasses five separate marine protected areas accessible from six of the eight main Hawaiian Islands. These areas support more than half of the North Pacific humpback whale population. Researchers working at Hawaiian Islands Humpback Whale National Marine Sanctuary partner with institutions and universities to better understand the whales and how to best manage their habitat. From December to March, visitors to the sanctuary can see whales from shore and at sea, and participate in the Sanctuary Ocean Count, a citizen science program designed to help track the number of whales visiting Hawai'i.

While these majestic creatures are a big part of the sanctuary, the marine life is not limited to humpback whales. The waters are also home to endangered Hawaiian monk seals, seabirds and shorebirds, dolphins, sea turtles, and a variety of bottom-dwelling fish, eels, octopus, squid, and crustaceans that crowd around the vibrant corals.

Mallows Bay–Potomac River National Marine Sanctuary

Designated: *September 3, 2019*
Size: *18 square miles*
Location: *Potomac River, Maryland*

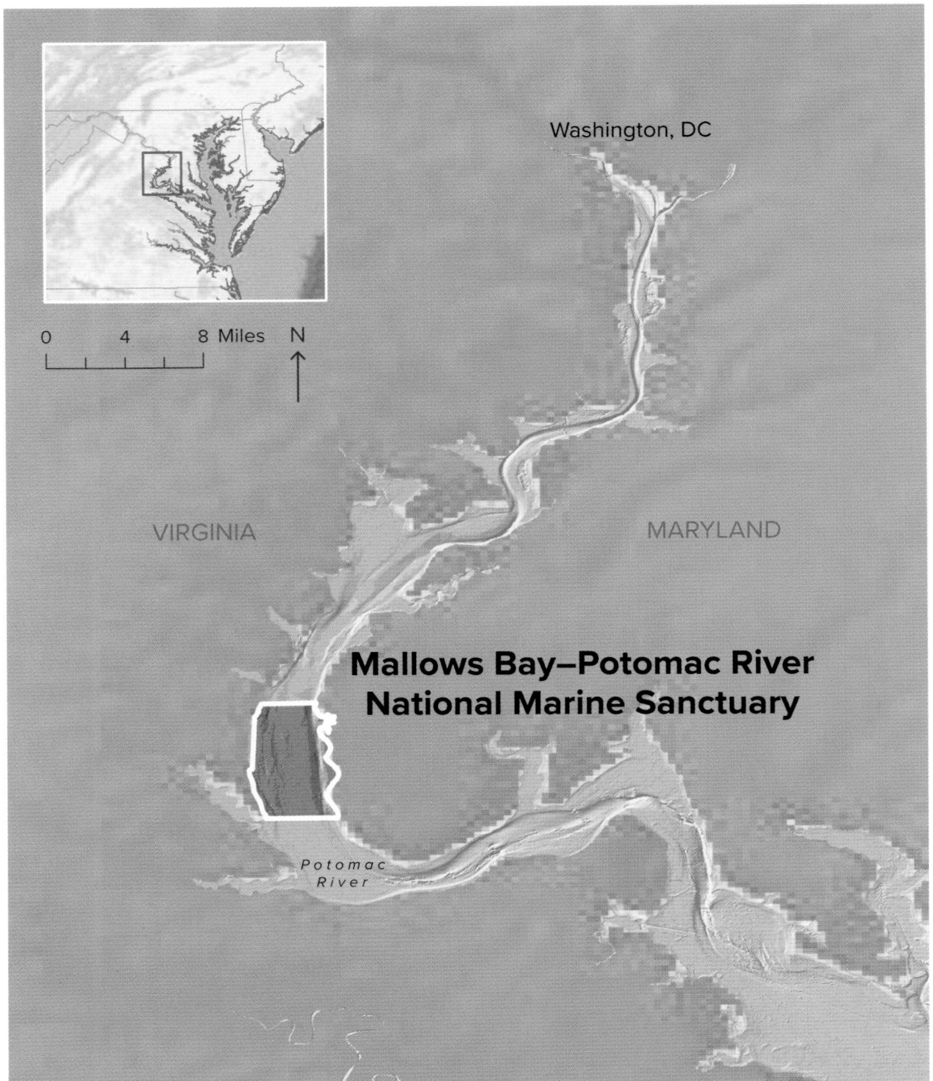

Forty miles outside Washington, DC, lies Mallows Bay–Potomac River National Marine Sanctuary. The sanctuary's rich history and heritage date back nearly 12,000 years and include the traditional homelands of the Piscataway Indian Nation, Piscataway Conoy Confederacy and Sub-Tribes of Maryland, and the Patawomeck Indian Tribe.

Mallows Bay and its surrounding waters comprise the largest wooden ship graveyard in the Western Hemisphere. This 18-square-mile sanctuary boasts a diverse collection of historic shipwrecks dating back to the Revolutionary War, but it is most renowned for the remains of over 100 wooden steamships known as the "Ghost Fleet." The ships, built between 1917 and 1919, were part of America's engagement in World War I. Their construction at more than 40 shipyards in 17 states reflected a massive wartime effort that drove economic development of communities and related maritime services as well as the expansion of the merchant marines. The ships' construction made the United States, for the first time in history, the greatest shipbuilding nation in the world.

The war ended before the ships could be used, and many of them were scuttled in the Potomac River for the purpose of salvaging scrap metal. Nearly a century of natural processes has gradually transformed these ships into ecologically valuable habitats. The overgrown wrecks now form a series of distinctive islands, intertidal habitats, and underwater structures critical to fish, beaver, and birds such as osprey, blue heron, and bald eagles. The extraordinary blending of history and ecology attracts and captivates visitors.

A kayaker explores the wrecks of Mallows Bay's Ghost Fleet.

The Ghost Fleet is your gateway to recreation and educational opportunities, whether you choose to walk the shore, paddle through history, or fish its waters. It emerges at low tide to reveal both vessel construction and the lasting impacts of ship breaking and environmental alteration. Kayakers can follow an interpretive water trail or steer their own course through the relative solitude of these waters.

Monitor National Marine Sanctuary

Designated: *January 30, 1975*
Size: *1 square mile*
Location: *North Carolina*

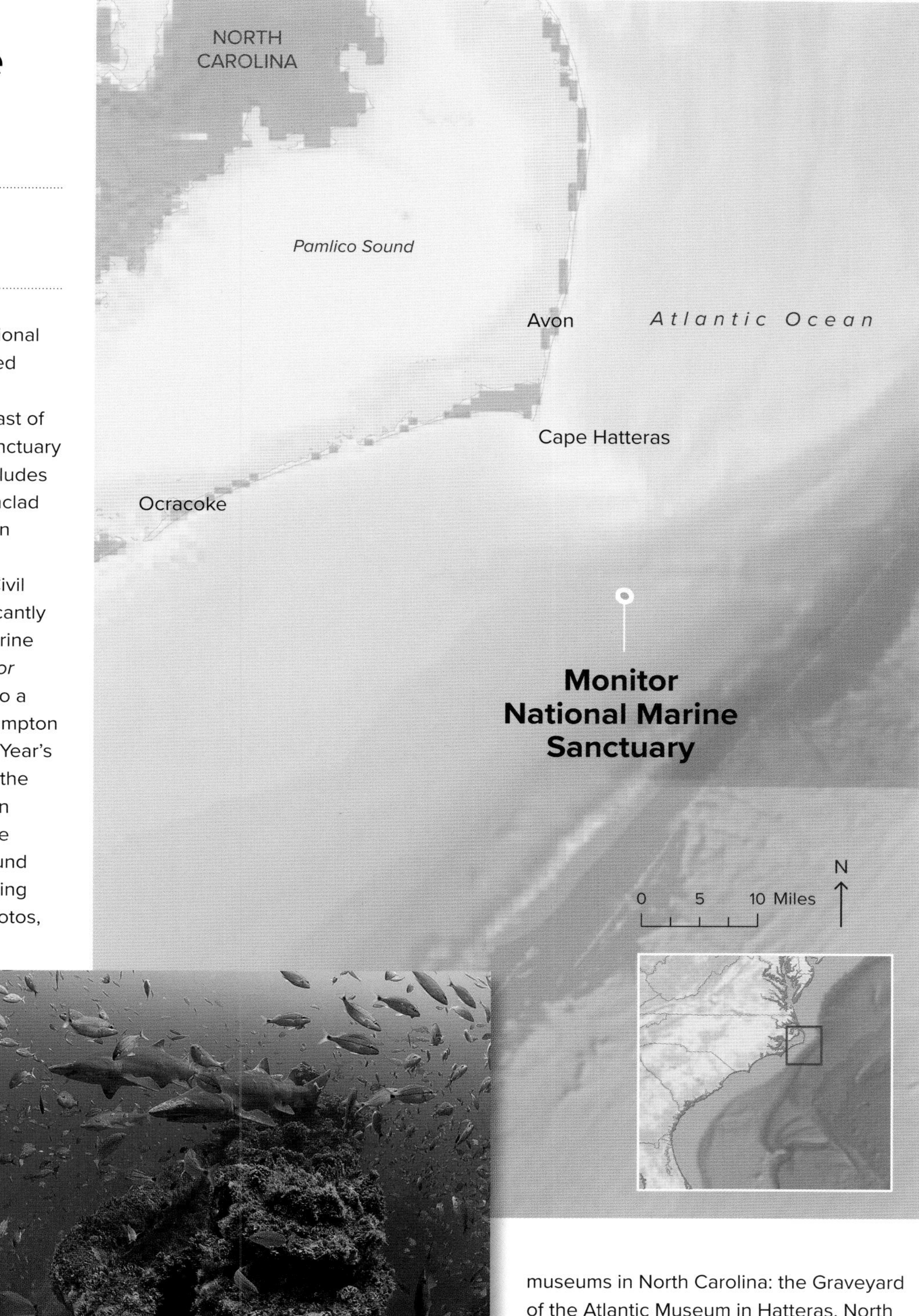

On January 30, 1975, Monitor National Marine Sanctuary was designated as our nation's first national marine sanctuary. Located 16 miles off the coast of Cape Hatteras, North Carolina, the sanctuary protects an area of open seas that includes the wreck of the USS *Monitor*, the ironclad warship launched by the Union Navy in February 1862.

As the prototype for a class of US Civil War ironclad warships, *Monitor* significantly altered both naval technology and marine architecture in the 19th century. *Monitor* fought the Confederate CSS *Virginia* to a draw in March 1862 in the Battle of Hampton Roads. Less than a year later, on New Year's Eve, *Monitor* was lost in a storm; 16 of the men aboard did not survive. In 1973, an oceanographic expedition studying the continental shelf off Cape Hatteras found a ship's remains. After months of imaging the area and reviewing videos and photos, it was finally determined that the unique-shaped hull of the *Monitor* had finally been found.

Dolphins, manta rays, sand tiger sharks, and black sea bass are just a few of the oceangoing fish that swim in the sanctuary's waters. NOAA and partners, including World War II veterans of the merchant mariners, are working to expand the sanctuary to an area known as the "Graveyard of the Atlantic," where the wrecks of steamships, freighters, tankers, and U-boats—consumed by war, piracy, and storms—lie below the surface. Some wrecks of the "Graveyard of the Atlantic" can be seen from shore, by snorkeling or by scuba diving, but *Monitor* lies about 230 feet deep in cool, rough waters, making it inaccessible to all but very experienced deep-ocean divers. Visitors can immerse themselves in *Monitor*'s history at the sanctuary's official visitor center; the Mariners' Museum in Newport News, Virginia; or at three partner museums in North Carolina: the Graveyard of the Atlantic Museum in Hatteras, North Carolina Aquarium on Roanoke Island, and the North Carolina Maritime Museum in Beaufort. The Outer Banks Maritime Heritage Trail, which includes historic lighthouses, vistas of shipwrecks, and wildlife viewing opportunities, runs along Highway 12 from Nags Head to the southern tip of Hatteras Island.

Sharks keep watch over the wrecks near USS Monitor.

Monterey Bay National Marine Sanctuary

Designated: *September 18, 1992*
Size: *6,094 square miles*
Location: *Central California Coast*

Known as the "Serengeti of the Sea," Monterey Bay National Marine Sanctuary is biologically and culturally rich. The sanctuary preserves the Central California Coast's legacy of ocean trade and industry, as well as a spectacular seascape at least as vast and abundant as the great plains of Africa. But unlike the Serengeti, most of Monterey's splendors are hidden from human view.

Stretching offshore from Marin to Cambria for around 30 miles, and covering 276 miles of shoreline, Monterey Bay National Marine Sanctuary is one of the United States' largest national marine sanctuaries. In this place where rolling hills and steep, rocky cliffs meet sand dunes and the wild ocean, the land and sea combine to create a great variety of natural settings. The diversity of plant and animal life is high: sea otters drift among kelp forests, sea lions and elephant seals rest on sandy beaches, sea stars cling to rocks in the intertidal zone, sharks cruise in the offshore waters, colorful fish inhabit deep-sea coral communities, and seabirds circle and dive to feed.

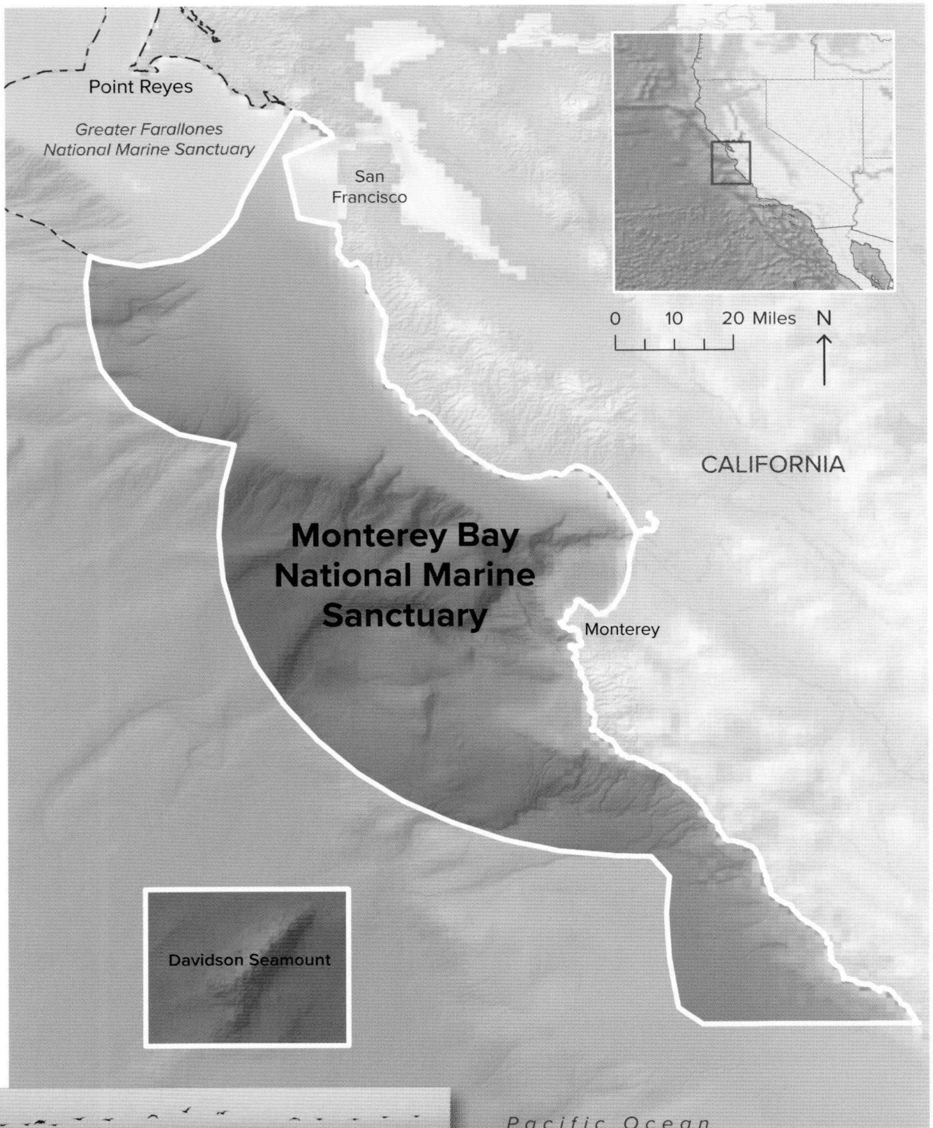

Extending 100 miles offshore, reaching depths of more than two and a half miles below the surface, lies the Monterey Canyon. This defining feature of Monterey Bay is one of North America's largest underwater canyons. If all its water were drained, the cliffs and valleys revealed would rival those of the Grand Canyon. Scientists are mapping the canyon using sophisticated instruments such as multibeam sonar, which beams a fan-shaped array of sound waves from a ship to the ocean bottom, to understand the ways this magnificent geological feature supports the ecosystem.

Find fabulous wildlife viewing and waves in Monterey Bay.

Visitors can dive in Whaler's Cove, surf in powerful and massive waves, kayak through kelp forests, and spot sea otters, whales, dolphins, and porpoises year-round. In spring and summer they can enjoy some of the best whale watching in the world, as gray, humpback, and blue whales are drawn to the sanctuary's water by the abundance of plankton. The Sanctuary Exploration Center in Santa Cruz and the Coastal Discovery Center in San Simeon offer interactive exhibits and displays. The Monterey Bay Aquarium in Monterey provides a distinctive window into the marine life of the bay.

Olympic Coast National Marine Sanctuary

Designated: *July 22, 1994*
Size: *3,188 square miles*
Location: *Washington's Outer Coast*

Off the rugged Olympic Peninsula, stretching along the coast and 25 to 50 miles into the sea, an array of sandy beaches, tide pools, rocky reefs, and deep-sea canyons create varied habitats for diverse marine life in Olympic Coast National Marine Sanctuary. Tufted puffins nest on rocky cliffs, and orcas hunt in open waters.

The sanctuary is a haven for 29 species of marine mammals, including toothed whales, baleen whales, and sea otters. Some live here year-round, like Steller and California sea lions, while others, like the gray whale, roam the open oceans and linger for a season in these protected waters as part of their annual migration. Below the water's surface, an otherworldly landscape includes communities of deep-sea corals and sponges.

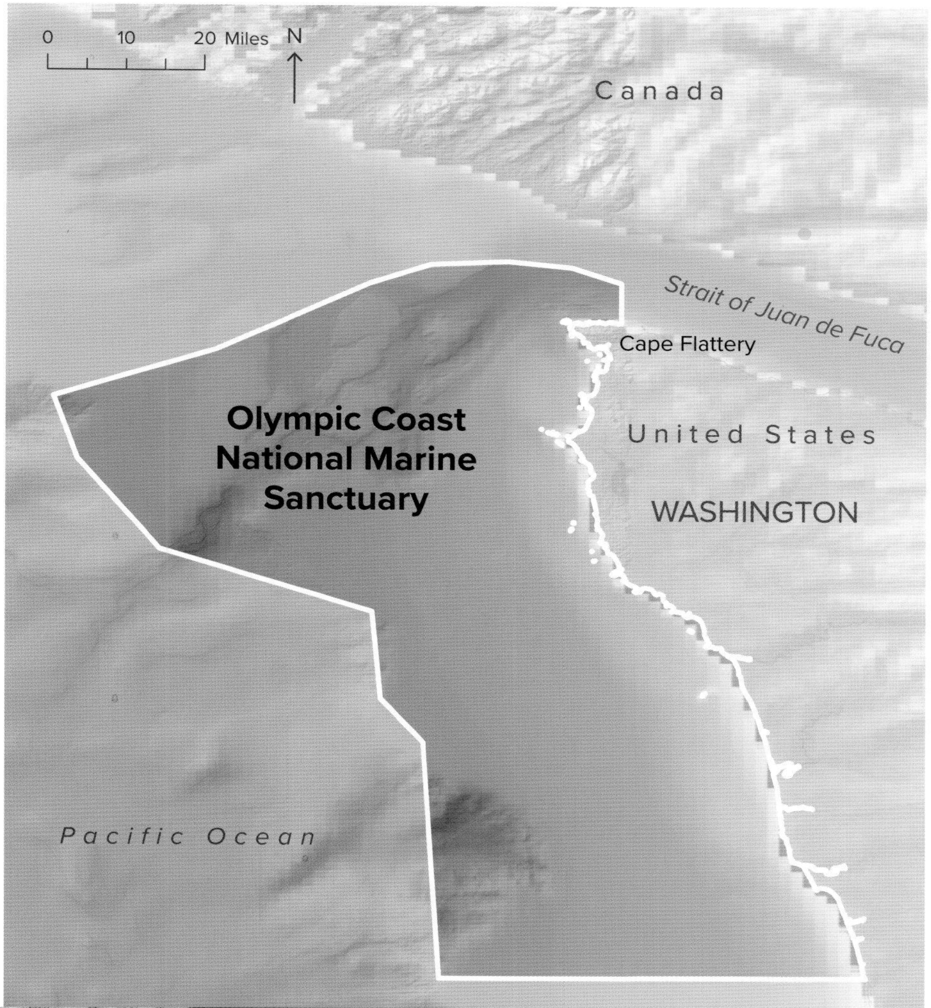

Three major canyons—Nitinat, Juan de Fuca, and Quinault—plunge from the land down to the sea, and then continue under the waves. In this sanctuary, wind-blown surface waters are replaced by cold, nutrient-rich water from below, creating a uniquely productive upwelling zone. The Olympic Coast's waters house dense kelp forests where fish and marine invertebrates thrive. Many types of fish, invertebrates, and seaweeds inhabit rocky reefs, while sea stars, hermit crabs, and anemones thrive in tide pools.

Olympic Coast is also defined by its long, rich history and cultural landscape. The Makah, Hoh, and Quileute tribes and the Quinault Indian Nation have lived here for thousands of years, drawing sustenance from the ocean and rivers. Human ties to this refuge are cherished and preserved in stories, songs, place names, and artifacts. The sanctuary works with other resource managers to evaluate ways to protect resources important to coastal communities, and explores the connections among wild lands, wildlife, and people.

Visitors can fish on charter boats, surf, scuba dive, and kayak among kelp forests. Other options include tide pooling, beachcombing, and watching wildlife. The Olympic Coast Discovery Center in Port Angeles, and annual events such as Makah Days, Grays Harbor Shorebird Festival, and the Seattle Aquarium's Family Science Weekend, offer opportunities for visitors to learn more; the sanctuary's Washington Coast Cleanup welcomes volunteers each year.

0 10 20 Miles N
Canada
Strait of Juan de Fuca
Cape Flattery
Olympic Coast National Marine Sanctuary
United States
WASHINGTON
Pacific Ocean

Anemones bloom like flowers in the Olympic Coast's waters.

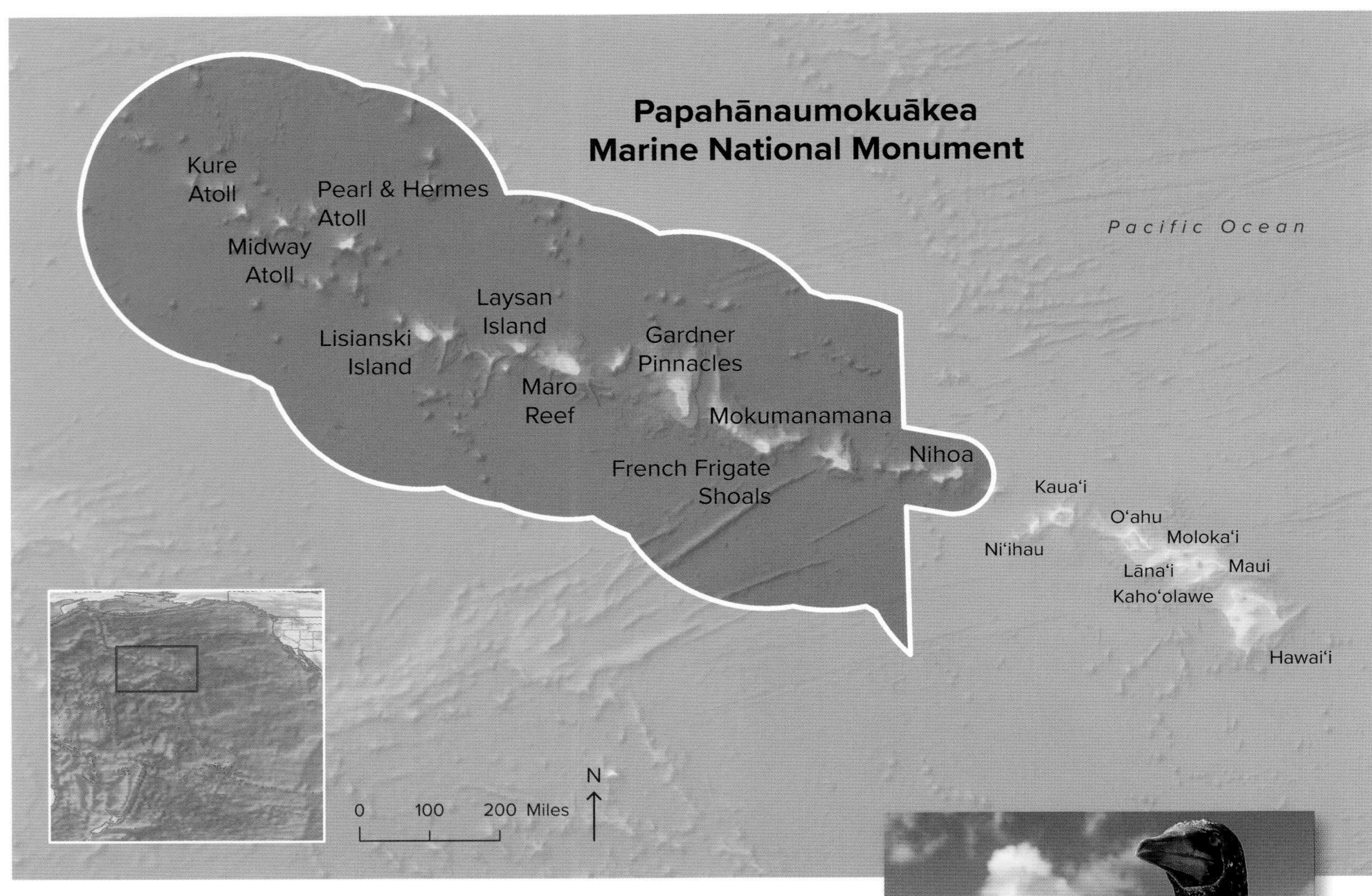

Papahānaumokuākea Marine National Monument

Designated: *June 15, 2006*
Size: *582,578 square miles*
Location: *Hawai'i*

The name Papahānaumokuākea commemorates the union of two Hawaiian ancestors: Papahānaumoku and Wākea. Together, they gave rise to the Hawaiian Archipelago, the taro plant, and the Hawaiian people. The marine monument encompasses 582,578 square miles of the vast Pacific Ocean and is bigger than all of the United States' national parks combined, making it the largest conservation area in the United States and one of the largest in the world. The remote location makes it one of the world's most untouched marine environments.

Coral islands, undersea volcanoes, flat-topped undersea mountains, banks, and shoals stretch for 1,350 miles to the northwest of the main Hawaiian Islands, creating a haven for marine life. Threatened green sea turtles and endangered Hawaiian monk seals are a just a few of the rare species that inhabit the island chain. Papahānaumokuākea is also home to 14 million seabirds, including the critically endangered Laysan duck, one of four bird species found only on these islands.

Papahānaumokuākea is a United Nations World Heritage Site, a designation given to places that are considered important to all of humankind. The monument helps preserve sacred places, stories, artifacts, and strong Polynesian cultural ties to the land and sea, dating back more than a thousand years.

Papahānaumokuākea is a refuge for rare birds.

Though the monument is not accessible to the public, there are many opportunities to learn about its natural and cultural treasures. The Mokupāpapa Discovery Center, Waikīkī Aquarium, and the Bishop Museum in Hawai'i exhibit artifacts from the monument; and the Nantucket Whaling Museum in Nantucket, Massachusetts, also holds artifacts from the region's whaling past.

Stellwagen Bank National Marine Sanctuary

Designated: *November 4, 1992*
Size: *842 square miles*
Location: *Massachusetts*

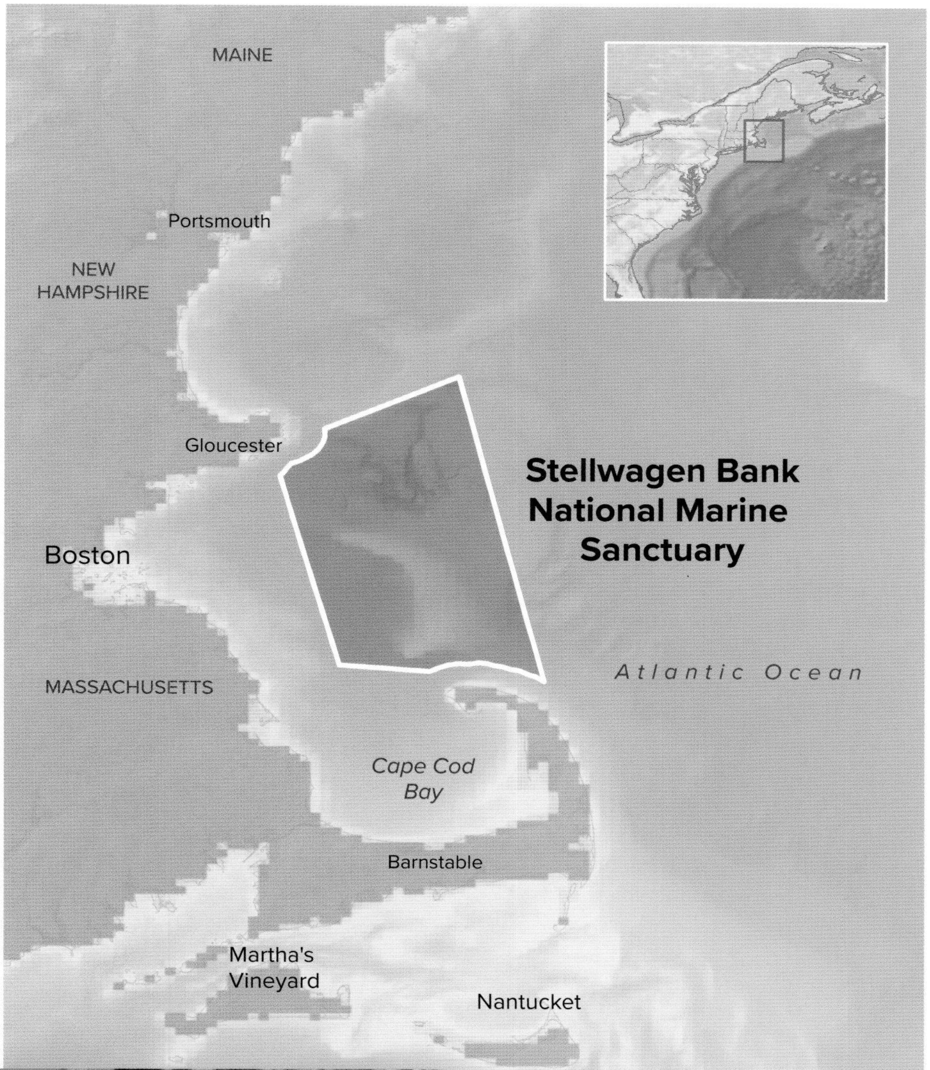

Each spring, hundreds of humpback whales make their way to the waters off Cape Cod. There, they hunt for a small fish known as the sand lance, gathering in spectacular feeding displays. Their feeding grounds are located in Gerry E. Studds Stellwagen Bank National Marine Sanctuary.

Stellwagen Bank formed as glaciers retreated during the last Ice Age, dropping sand, gravel, and boulders as they moved northward. As sea levels rose about 10,000 years ago, Stellwagen Bank slipped beneath the water. This rise is now the centerpiece of a diverse ecosystem named after Lieutenant Henry S. Stellwagen, US Navy, who surveyed the bank in the 1850s, and Gerry E. Studds, a representative in Congress for Massachusetts who fought for the site's designation as a sanctuary. As America's oldest commercial fishing grounds, the bank and its surrounding waters remain a destination for an active fishing fleet.

The sanctuary hosts one of the most biologically diverse ecosystems in the Gulf of Maine. In addition to humpback whales, fin whales and minke whales travel many miles to feed here; visitors can also see dolphins, porpoises, and seals. The sanctuary's rich waters serve as a stopover for large numbers of migrating birds, including shearwaters, storm petrels, fulmars, and gannets. Stellwagen Bank is one of the best places in the United States to see aquatic wildlife. Visitors to the sanctuary enjoy whale watching, birding, and other wildlife watching.

Historic New England shipping routes also cross the region, and over the course of centuries, the seafloor has become a repository for shipwrecks. Researchers have identified 47 wrecks in the sanctuary; there are undoubtedly many more.

The sanctuary encourages learning and exploration through its many programs, including an annual children's art contest and activities with local schools and libraries. Stellwagen Bank also focuses on research, whether it be tagging whales and seabirds to better understand feeding behaviors; analyzing the soundscape of the bank and surroundings; mapping the seafloor to delineate habitats and discover potential shipwreck sites; understanding wildlife, habitat, and cultural features; or studying historical ecology.

Whale-watching tours are popular at Stellwagen Bank.

Thunder Bay National Marine Sanctuary

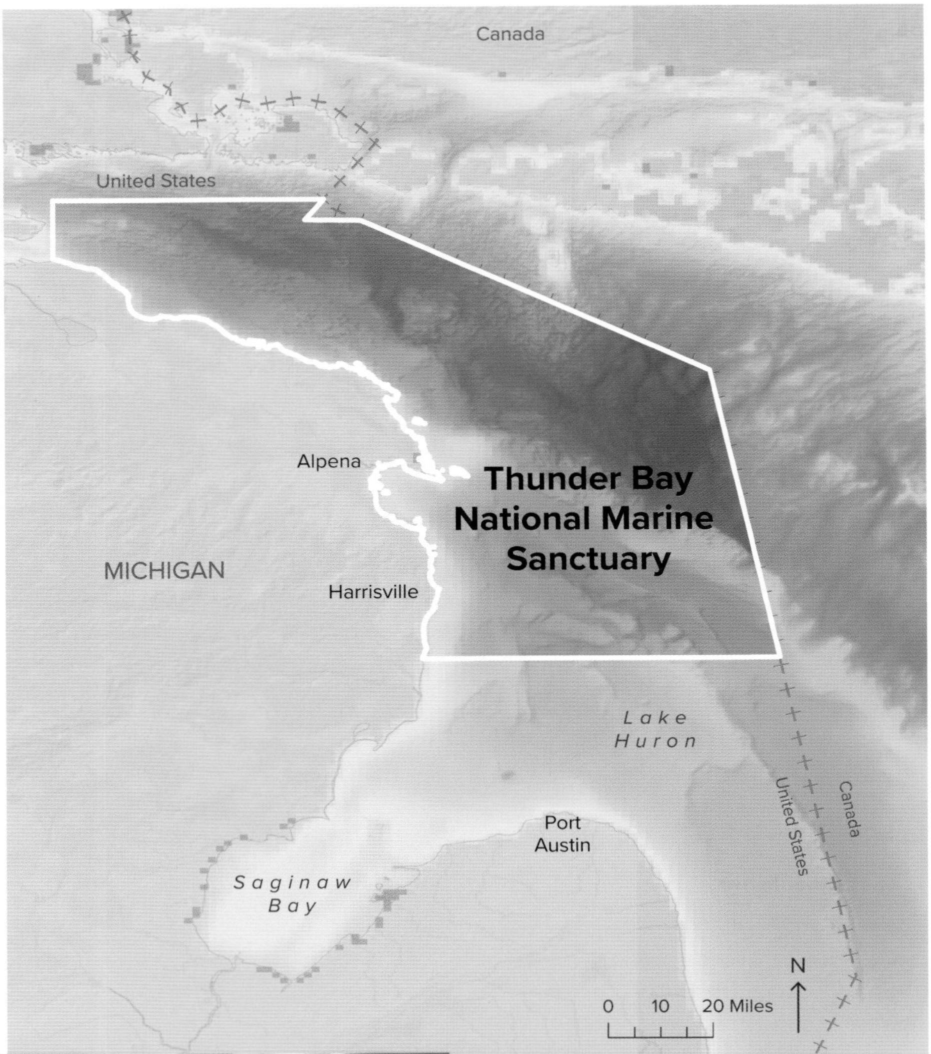

Designated: *September 25, 2000*
Size: *4,300 square miles*
Location: *Michigan*

For over 12,000 years, people have traveled on the Great Lakes. From Native American dugout canoes to wooden sailing craft and steel freighters, thousands of ships have made millions of voyages across the Great Lakes. In the northwestern part of Lake Huron and off the coast of Alpena, Michigan, lies "Shipwreck Alley," a perilous stretch of water in the Great Lakes. Unpredictable weather, murky fog banks, sudden gales, and rocky shoals claimed more than 200 ships; at least half these shipwrecks are found within Thunder Bay National Marine Sanctuary. Lake Huron's cold, fresh water ensures that Thunder Bay's shipwrecks are among the best preserved in the world. From an 1838 sidewheel steamer to a modern, 500-foot-long German freighter, the sanctuary's shipwrecks capture dramatic moments from centuries that transformed America. As a collection, they illuminate an era of enormous growth and remind us of risks taken and tragedies endured from the earliest explorations of the region through westward expansion in the 1800s, and up to modern lake traffic and trade.

Thunder Bay National Marine Sanctuary protects a nationally significant collection of shipwrecks and related maritime heritage resources that paddlers, divers, and snorkelers to the area can visit. Some, such as the wooden schooner *Portland* and the paddle-wheel steamer *Albany*, rest in shallow enough water that they can easily be explored from a paddleboard or a kayak. Other wrecks, like *Lucinda Van Valkenburg* and *W.P. Thew*, provide opportunities to new and experienced divers alike. Seasonal mooring buoys provide a safe attachment point for visiting boats, limiting damage to the shipwrecks.

Paddle boarding is a great way to enjoy Lake Huron.

Visitors to the sanctuary can also explore northeastern Michigan's maritime landscape, including life-saving stations, lighthouses, historic boats and ships, commercial fishing camps, docks, and working ports. The Great Lakes Maritime Heritage Center in Alpena, Michigan, with its immersive exhibits is the perfect gateway to get to know Thunder Bay.

Thunder Bay's shipwrecks are magnificent yet vulnerable; natural processes and human impacts threaten them. Through research, education, and community involvement, the sanctuary works to protect our nation's historic shipwrecks for future generations to experience.

HOW YOU CAN HELP

Make a Difference for Our Ocean and Great Lakes

Protect Biodiversity and Promote Climate Resiliency

National marine sanctuaries and marine national monuments safeguard biodiversity, preserve our cultural heritage, support a healthy ocean and Great Lakes, and promote resiliency to climate change. Nationally and internationally, scientists are calling for 30 percent of land and water to be protected by 2030. Do your part in helping preserve biodiversity by visiting, volunteering at, and supporting marine protected areas. Learn more about the National Marine Sanctuary Foundation's efforts to advocate for more protected areas in our ocean and Great Lakes, especially those that strongly safeguard natural and cultural resources, at marinesanctuary.org.

Eat Sustainable Seafood

Around the world, fisheries are threatened with collapse due to unsustainable fishing methods and ecosystem destruction. As fisheries near shore are depleted, fishing fleets travel greater distances for catches. These trips require more labor to meet the demand for fish and are fueling human rights abuses and trafficking. When grocery shopping or dining out, choose seafood that is sustainable. Get the Seafood Watch app from the Monterey Bay Aquarium, check for the Marine Stewardship Council label on seafood, and ask your local restaurants and markets to buy from sustainable fisheries and local fishers. Urge policymakers and legislators to combat illegal, unreported, and unregulated practices and to enforce laws against human trafficking in fisheries.

Reduce Carbon Emissions and Your Energy Use

Climate change impacts on our ocean and Great Lakes include increasing ocean temperatures, changing ocean currents, sea level rise, acidification of the ocean, and more extreme weather events. Encourage international and national action to reduce carbon emissions, and support businesses that are taking voluntary action to address climate change. Individually, take simple actions to reduce your own energy use. Leave your car at home when you can and ride a bike, walk, or use public transportation. Use high-efficiency appliances and light bulbs in your home. Turn off and unplug appliances and electronics when they aren't in use. Turn up your thermostat a few degrees in the summer and down a few degrees in the winter.

Use Less Plastic and Help Clean Up Marine Debris

Plastics that end up in waterways as pollution contribute to habitat destruction, and entangle and kill many marine animals. Reduce, reuse, and recycle! Buy products with less packaging. Use reusable water bottles, paper straws (if you need one), and cloth grocery bags. Recycle whenever possible. Don't release helium balloons outside because, once they end up in waterways, they harm animals that mistake them for food. Join local efforts to clean up coasts and waterways, such as rivers and streams. Support programs at national marine sanctuaries that work with local communities and businesses to clean up marine debris both on the coast and under the water.

Respect Ocean Wildlife and Habitats

Whether you're exploring the ocean and Great Lakes by diving, surfing, boating, or by relaxing on the beach, be sure to show your appreciation and clean up after yourself. Keep your distance from seabirds, mammals, and other wildlife so that you don't disturb their feeding or nesting grounds. Tread lightly on tide pools and other shore habitats. Don't remove rocks or corals.

Dispose of Household and Hazardous Materials Properly

Motor oil, antifreeze, and other hazardous materials harm our oceans when they aren't disposed of properly. Dispose of hazardous waste in an environmentally safe way. Fix car leaks, and recycle motor oil. Don't flush pharmaceuticals, nondegradable products, or cleaning products with bleach or phosphates down the drain.

Use Less Fertilizer

When fertilizers are used in gardening and agriculture, the excess eventually ends up in waterways, where it can create a "dead zone" with little or no oxygen to support ocean life. Use fertilizer sparingly, and use ecological or organic fertilizers in your gardens and on your lawns when possible.

Combat Invasive Species

An invasive species is one that is not native to an ecosystem and that, once introduced, causes environmental and economic harm or endangers human health. Invasive species are one of the leading threats to biodiversity. Never release a pet into the wild. Support efforts to establish strict standards to treat ballast water in ships. If you are a boater, clean, drain, and dry your boat after each use. Participate in local efforts to monitor and eradicate invasive species.

ACKNOWLEDGMENTS

Many people made *America's Marine Sanctuaries* possible, and the National Marine Sanctuary Foundation is grateful to all of them. The time they dedicated to creating this volume and sharing compelling stories is a testament to the importance of the National Marine Sanctuary System to our one global ocean. We would especially like to thank Elizabeth Moore for sharing her vision for the book with us and inspiring this effort. And most notably, we are thankful for all the dedicated staff and volunteers who work to protect our national marine sanctuaries for current and future generations:

David Alberg, Allison Alexander, Pelika Andrade, John Armor, Carol Bernthal, Reed Bohne, Allie Braun, Matt Brookhart, Maria Brown, Kathy Broughton, Athline Clark, Shannon Colbert, Pete DeCola, Andrew DeVogelaere, James Delgado, Mimi D'Iorio, Beth Dieveney, Bill Douros, Kelly Drinnen, Sarah Fangman, Lilli Ferguson, Claire Fackler, Megan Forbes, Steve Gittings, Kevin Grant, Jeff Gray, Tracy Hajduk, Ben Haskell, Melanie Herrera, Becky Holyoke, Kristina Kekuewa, Carol King, Francesca Koe, Susan Langley, Bob Leeworthy, Kirsten Lindquist, Danielle Lipski, Steve Lonhart, Ed Lyman, Hannah MacDonald, Catherine Marzin, Paul Michel, Chris Mobley, Enti Mooskin, Michael Murray, Apulu Veronika Molio'o Mata'utia Mortenson, Seaberry Nachbar, Paul Orlando, Jennifer Ortiz, Atuatasi-Lelei Peau, Naomi Pollack, Kevin Powers, Ben Prueitt, Kalani Quiocho, Marcus Reamer, Stan Rogers, Jan Roletto, Gabriela Sarri-Tobar, G.P. Schmahl, Mary Jane Schramm, Danielle Schwarzmann, George Sedberry, Kate Spidalieri, Jennifer Stock, Matt Stout, Lisa Symonds, Mitchell Tartt, Kate Thompson, Allen Tom, Liz Weinberg, Chip Weiskotten, and David Wiley.

The wonders of the National Marine Sanctuary System are hard to describe in words. Thank you to all the photographers whose images of the sanctuaries brought their ineffable beauty alive:

Robin Agarwal, Jessie Altstatt, Pelika Andrade, Na'alehu Anthony, Steve Bauer, Melody Bentz, Tasia Blough, Bob Bonde, Alisia Carlson, Tane Casserley, Wendy Cover, Douglas Croft, N. Davis, Brian Dort, Tiffany Doug, Daryl Duda, Keith Ellenbogen, Claire Fackler, Keith Flood, Peter Flood, Ari Friedlaender, Stephen Frink, Stephen Gnam, Chuck Graham, Andrew Gray, Kimberly Hernandez, Joe Hoyt, Katy Laveck Foster, Matthew Lawrence, Robert Lee, Beata Lerman, Ed Lyman, Koa Matsuoka, Paulo Maurin, Greg McFall, Matt McIntosh, J. Moore, Eric Palmer, David Ruck, Anne Mary Schaefer, G.P. Schmahl, Stephen Sellers, Robert Schwemmer, Ian Shive, Anastasia Strebkova, Bruce Sudweeks, Mark Sullivan, Ryan Tabata, Bob Talbot, Nicol Uibel, Shawn Verne, Matt Vieta, Sophie Webb, Liz Weinberg, Olivia Williamson, Nick Zachar, and the NOAA Office of Ocean Exploration and Research.

REFERENCES

Berkes, Fikret. *Sacred Ecology.* 3rd ed. Routledge, 2012.

A Bibliographic Listing of Coastal and Marine Protected Areas: A Global Survey. WHOI Technical Report, 1986.

Brax, Jeff, et al. "Zoning the Ocean." *Ecology Law Quarterly* 29, no. 1 (2002): 71–130.

Brockington, Dan, Rosaleen Duffy, and Jim Igoe. *Nature Unbound: Conservation, Capitalism and the Future of Protected Areas.* Earthscan, 2008.

Congressional Record 17:14 (June 9, 1971) pp. H19057–58.

Costanza, Robert. "The Ecological, Economic, and Social Importance of the Ocean." *Ecological Economics,* 1999.

Craig, Steve. *Sports and Games of the Ancients.* Greenwood Press, 2002.

Davenport, D., J. Johnson & J. Timbrook. "The Chumash and the Swordfish." *Antiquity* 67, no. 255 (1993): 257–72.

"The Drive to Save America's Shorelines." *U.S. News and World Report,* July 31, 1972.

"The Dying Ocean." *Time*, September 28, 1970.

Effective Use of the Sea. Report of the Panel on Oceanography of the President's Science Advisory Committee. Washington, DC: US Government Printing Office, 1966. worldcat.org/title/effective-use-of-the-sea-report-of-the-panel-on-oceanography-of-the-presidents-science-advisory-committee/oclc/60038057&referer=brief_results.

Epting, John. "National Marine Sanctuary Program: Balancing Resource Protection with Multiple Use." *Houston Law Review,* 1980–81.

Forestell, Paul H. "Popular Culture and Literature." In *Encyclopedia of Marine Mammals*, 2nd ed., edited by W.F. Perrin, B. Würsig, and J.G.M. Thewissen. Elsevier, 2009.

"Framework for the National System of Marine Protected Areas of the United States of America." Silver Spring, MD:

National Marine Protected Areas Center, March 2015.

Gauld, George. *Observations on the Florida Kays, Reef and Gulf*. W. Faden, 1796.

Gifford-Gonzalez, D., S.D. Newsome, P.L. Koch, T.P. Guilderson, J.J. Snodgrass, and R.K. Burton. “Archaeofaunal Insights on Pinniped-Human Interactions in the Northeastern Pacific.” In *The Exploitation and Cultural Importance of Sea Mammals*, edited by G.G. Monks. Oxbow Books, 2002.

Gomez-Pompa, Antonia, and Andrea Kaus. “Taming the Wilderness Myth.” *Bioscience,* 1992.

Hastings, J., et al. “Safeguarding the Blue Planet: Six Strategies for Accelerating Ocean Protection.” *Parks* 18*,* no. 13 (2012).

Herman, Brother, coll. and trans. *Tales of Ancient Samoa*. 1966.

International Union for the Conservation of Nature; iucn.org.

Ivanovici, A., ed. *Inventory of Declared Marine and Estuarine Protected Areas in Australian Waters*, Spec. Publ. 12, Australian National Parks and Wildlife Service. Canberra, Australia, 1984.

Jackson, E.L., ed. *Constraints to Leisure*. Venture Publishing, 2005.

Jensen, A.M., G.W. Sheehan, and S.A. MacLean. “Inuit and Marine Mammals.” In *Encyclopedia of Marine Mammals,* 2nd ed., edited by W.F. Perrin, B. Würsig, and J.G.M. Thewissen. Elsevier, 2009.

Johannes, R.E. “Traditional Marine Conservation Methods in Oceania and Their Demise.” *Annual Review of Ecology and Systematics,* 1978.

Johnson, L.L. “Aleut Sea Mammal Hunting: Ethnohistorical and Archaeological Evidence.” In *The Exploitation and Cultural Importance of Sea Mammals*, edited by G.G. Monks. Oxbow Books, 2002.

Jones, P.J.S., Qiu, Wanfei & Elizabeth De Santo. “Governing Marine Protected Areas: Social–Ecological Resilience Through Institutional Diversity.” *Marine Policy* 41 (2013): 5–13.

Kramer, Augustin. *The Samoa Islands: An Outline of a Monograph with Particular Attention of German Samoa*. 1901. Translated by Theodore Verhaaren. University of Hawaii Press, 1994.

Library of Liberty; http://oll.libertyfund.org/pages/1641-massachusetts-body-of-liberties.

Lubchenco, J., and Steven D. Gaines. “A New Narrative for the Ocean.” *Science* 364, no. 6444 (June 7, 2019): 911. DOI: 10.1126/science.aay2241.

Merrell, W., M. Katsouros, and J. Bienski. “The Stratton Commission: The Model for a Sea Change in National Marine Policy.” *Oceanography* 14 (2001).

National Academy of Public Administration. *Protecting Our National Marine Sanctuaries*. Washington, DC: NAPA, 2000.

National Marine Protected Areas Center; marineprotectedareas.noaa.gov.

National Oceanic and Atmospheric Administration; noaa.gov.

National Workshop on Sanctuaries (Washington, DC). Proceedings of *Marine and Estuarine Sanctuaries*. Special scientific report no. 70. Virginia Institute of Marine Science, College of William and Mary, 1973. dx.doi.org/doi:10.21220/m2-yj3e-ne58.

Neufield, David. “Indigenous Peoples and Protected Heritage Areas: Acknowledging Cultural Pluralism.” In *Transforming Parks and Protected Areas,* edited by Kevin Hannah, Douglas Clark, and Scott Slocombe. Routledge, 2008.

Oregon State University, IUCN World Commission on Protected Areas, Marine Conservation Institute, National Geographic Society, and UNEP World Conservation Monitoring Centre. “An Introduction to the MPA Guide.” 2019. protectedplanet.net/c/mpa-guide.

“Parks and Peoples: The Social Impacts of Protected Areas.” *Annual Review of Anthropology* 35 (2006).

Polefka, Shiva, and Billy DeMaio. *The Dividends of Coastal Conservation in the United States: An Economic Analysis of Coastal and Ocean Parks*. Washington, DC: Center for American Progress, 2016.

Policy Study of Marine and Estuarine Sanctuaries. Gloucester Point, VA: Virginia Institute of Marine Science, 1973.

Proceedings of the National Conference on Outdoor Recreation. Washington, DC: US Government Printing Office, 1924.

Randall, John E. *Conservation in the Sea: A Survey of Marine Parks*. Oryx, 1969.

Ratified Treaty No. 286, Documents Relating to the Negotiation of the Treaty of January 31, 1855 with the Makah Indians. http://digicoll.library.wisc.edu/cgi-bin/History/History-idx?type=article&did=History.IT1855no286.i0001&id=History.IT1855no286&isize=M.

Ready to Perform? Planning and Management at the National Marine Sanctuary Program. Washington, D.C.: National Academy of Public Administration, 1993. napawash.org/uploads/Academy_Studies/06-11.pdf.

“Representativeness of Marine Protected Areas of the United States.” Silver Spring, MD: National Marine Protected Areas Center, January 2015.

Revelle, Roger. “A Long View from the Beach.” *New Scientist,* 1964.

Roland, Alex, et al. *The Way of the Ship: America’s Maritime History Reenvisioned, 1600 to 2000*. John Wiley and Sons, Inc., 2007.

Shepherd, James F., and Gary M. Walton. *Shipping, Maritime Trade and the Economic Development of Colonial North America*. Cambridge University Press, 1972.

Spalding, H.L., K.Y. Conklin, C.M. Smith, C.J. O’Kell, and A.R. Sherwood. “New Ulvaceae (Ulvophyceae, Chlorophyta) from Mesophotic Ecosystems Across the Hawaiian Archipelago.” *Journal of Psychology*, 2016.

Stellwagen Bank Marine Historical Ecology Final Report. 2010. Marine Sanctuaries Conservation Series ONMS-10-04. https://nmssanctuaries.blob.core.windows.net/sanctuaries-prod/media/archive/science/conservation/pdfs/sb_historical.pdf.

Tell, William. “Marine Sanctuaries: Balancing of Energy vs. Environmental Needs.” *Natural Resources Law* 6 (1973).

Topal, R.S., and A. Ongen. “Integrate Protected Areas into Broader Land, Seascapes, and Sectors from a Social Responsibility of the Environment Perspective.” *Social Responsibility Journal* 2, no. 3/4 (2006): 291–99.

Turner, George. *Samoa, a Hundred Years Ago and Long Before*. Macmillan, 1884. https://books.google.com/books/about/Samoa_a_Hundred_Years_Ago_and_Long_Befor.html?id=itaG4-8bglUC.

US Department of the Interior; doi.gov.

INDEX

Page numbers in *italics* refer to figures.

IMAGE CREDITS

2–3: Matt McIntosh/NOAA; 6–7: Ed Lyman/NOAA permit #14682; 9: Mark Sullivan/NOAA; 10–11: Stephen Gnam; 13: Robin Agarwal; 14–15: reproduced from the GEBCO world map 2014, www.gebco.net; 15: National Marine Sanctuary Foundation; 16: Daryl Duda; 17: Koa Matsuoka; 18–19: Bruce Sudweeks; 21: NASA; 23: Shawn Verne; 24: NOAA; 25: Olivia Williamson; 26–27: Ian Shive/Tandem; 28: Greg McFall/NOAA; 30–31: Douglas Croft; 33: Wendy Cover/NOAA; 34: J. Moore/NOAA permit #15240; 37: Andrew Gray/NOAA; 39: Matt McIntosh/NOAA; 42: Greg McFall/NOAA; 44–45: Douglas Croft; 47: Beata Lerman; 49: Scott Bauer; 51: Greg McFall/NOAA; 53: Peter Flood; 55: Keith Flood; 58: Ian Shive/Tandem; 59: Ed Lyman/NOAA; 60, 61: NOAA; 62: T. Blough/NOAA (MMHSRP permit #18786-02); 63: Ed Lyman/NOAA permit #14682; 64–65: Nick Zachar/NOAA; 66: NOAA Office of Ocean Exploration and Research; 67: Paulo Maurin/NOAA; 68–69: Greg McFall/NOAA; 70: Nick Zachar/NOAA; 71: Northwest Fisheries Science Center; 72: Liz Weinberg/NOAA; 73: David Ruck/NOAA; 74–75: Katy Laveck Foster; 76: Sophie Webb/NOAA/Point Blue; 77: Robert Lee/NOAA; 78–79: Matt Vieta/NOAA; 80: Matt McIntosh/NOAA; 81: Anne Mary Schaefer; 82, 83: NOAA; 84: Douglas Croft; 85: Robert Schwemmer/NOAA; 86, 87: Keith Flood; 88–89: Douglas Croft; 90: Chuck Graham; 91: Bruce Sudweeks; 92–93: Beata Lerman; 94: G.P. Schmahl/NOAA; 95, 96: Steve Bauer; 97: G.P. Schmahl/NOAA; 98: Claire Fackler/NOAA; 99: Bob Bonde/USGS; 100–101: Stephen Frink; 102–105: Greg McFall/NOAA; 106: Bruce Sudweeks; 107: NOAA; 108–109: Stephen Sellers/NOAA; 110, 111: Matt McIntosh/NOAA; 112: Kimberly Hernandez; 113: NOAA; 114: © Keith Ellenbogen; 115: NOAA/NOAA permit #14245; 116–117: ©Keith Ellenbogen; 118: Brian Dort; 119: Tane Casserley/NOAA; 120–121: Tane Casserley/NOAA; 122–123: Bryan Dort; 125: David Ruck/NOAA; 126–127: Matt McIntosh/NOAA; 128: Robert Schwemmer/NOAA; 130: Na'alehu Anthony/Polynesian Voyaging Society; 132: David Ruck/NOAA; 133: Robert Schwemmer/NOAA; 134–135: Matt McIntosh/NOAA; 136: Matthew Lawrence/NOAA; 139: Matt McIntosh/NOAA; 140–141, 143: Tane Casserley/NOAA; 144: Greg McFall/NOAA ; 146–147: © Keith Ellenbogen; 149: Olivia Williamson; 151: Eric Palmer/NOAA; 152–153, 155: Matt McIntosh/NOAA; 157: NOAA; 158–159: Robert Schwemmer/NOAA; 161: Alisia Carlson/Dive Shop at Ocean Reef Club; 162–163, 164–165: Matt McIntosh/NOAA; 167: Tiffany Doug; 168: Nicol Uibel/NOAA; 169: Pelika Andrade; 170–171: Matt McIntosh/NOAA; 172: Anastasia Strebkova; 173: David Ruck/NOAA; 175: Tasia Blough/NOAA; 176: Nick Zachar/NOAA; 177: Jessie Altstatt/NOAA; 178–179: Tane Casserley/NOAA; 181: Nick Zachar/NOAA; 183: G.P. Schmahl/NOAA; 184: Stephen Frink; 186: NOAA; 187: N. Davis/NOAA permit #932-1905; 188–189: Ryan Tabata/NOAA; 190: Melody Bentz; 192: Bruce Sudweeks; 194–195: Keith Ellenbogen; 196: Robin Agarwal; 198–199: David Ruck/NOAA; 202t: NOAA; 202b: Matt McIntosh/NOAA; 203t: NOAA; 203b: Matt McIntosh/NOAA; 204t: NOAA; 204b: Joe Hoyt/NOAA; 205t: NOAA; 205b: Stephen Frink; 206t: NOAA; 206b: G.P. Schmahl/NOAA; 207t: NOAA; 207b: Greg McFall/NOAA; 208t: NOAA; 208b: Matt McIntosh/NOAA; 209t: NOAA; 209b: Matt McIntosh/NOAA; 210t: NOAA; 210b: Matt McIntosh/NOAA; 211t: NOAA; 211b: Joe Hoyt/NOAA; 212t: NOAA; 212b: Nick Zachar/NOAA; 213t: NOAA; 213b: Katy Laveck Foster; 214t: NOAA; 214b: Andrew Gray/NOAA; 215t: NOAA; 215b: Ari Friedlaender/NOAA; 216t: NOAA; 216b: Bryan Dort; 224: Stephen Frink.

Christ of the Abyss is located in the waters near Key Largo in Florida Keys National Marine Sanctuary. In addition to attracting numerous invertebrates that have attached to its surface, giving it a colorful, living texture, this nine-foot-tall bronze statue is also a popular destination for snorkelers and divers.